KT-467-691

Contents

Acknowledgements

I should like to thank Victor Lee for his encouragement in the preparation of this book. My first conversations about Robert Frost's poetry were with Joseph Treasure to whom I owe a great deal – both as a colleague and a friend.

The text used for poems by Robert Frost in this edition is taken from *The Poetry of Robert Frost*, edited by Edward Connery Lathem. We are grateful to Random House for permission to reproduce them.

We gratefully acknowledge permission to reproduce 'Easter Monday' by Eleanor Farjeon from *The Last Four Years* (OUP, 1958) by permission of David Higham Associates.

Editors

Dr Victor Lee, the series editor, read English at University College, Cardiff. He was later awarded his doctorate at the University of Oxford. He has taught at secondary and tertiary level, and worked for twenty-seven years at the Open University. Victor Lee's experience as an examiner is very wide: he has been a Chief Examiner in English A-level for three different boards stretching over a period of almost thirty years.

Adrian Barlow, read English at University College, Durham. He was formerly Head of English and Director of Studies at Monmouth School, and is now an education consultant and a Chief Examiner in A-level English Literature. His publications include *The Calling of Kindred* (1993) and *Six Poets of the Great War* (1995).

Foreword

Oxford Student Texts are specifically aimed at presenting poetry and drama to an audience which is studying English Literature at an advanced level. Each text is designed as an integrated whole consisting of three main parts. The poetry or the play is placed first to stress its importance and to encourage students to enjoy it without secondary critical material of any kind. When help is needed on other occasions, the second and third parts of these texts, the Notes and the Approaches, provide it.

The Notes perform two functions. First, they provide information and explain allusions. Secondly, and this is where they differ from most texts at this level, they often raise questions of central concern to the interpretation of the poem or the play being dealt with, particularly in the use of a general note placed at the beginning of the particular notes.

The third part, the Approaches section, deals with major issues of response to the particular selection of poetry or drama, as opposed to the work of the writer as a whole. One of the major aims of this part of the text is to emphasize that there is no one right answer or interpretation, but a series of approaches. Readers are given guidance as to what counts as evidence, but, in the end, left to make up their mind as to which are the most suitable interpretations, or to add their own.

To help achieve this, the Approaches section contains a number of activity-discussion sequences, although it must be stressed that these are optional. Significant issues about the poem or the play are raised in these activities. Readers are invited to tackle these activities before proceeding to the discussion section where possible responses to the questions raised in the activities are considered. Their main function is to engage readers actively in the ideas of the text. However, these activity-discussion sequences are so arranged that, if readers wish to treat the Approaches as continuous prose and not attempt the activities, they can.

At the end of each text there is also a list of Tasks. Whereas the activity-discussion sequences are aimed at increasing understanding of the literary work itself, these tasks are intended to help explore ideas about the poetry or the play after the student has completed the reading of the work and the studying of the Notes and Approaches. These tasks are particularly helpful for coursework projects or in preparing for an examination.

Victor Lee *Series Editor*

Foreword

The Poems

The Pasture

I'm going out to clean the pasture spring;
I'll only stop to rake the leaves away
(And wait to watch the water clear, I may):
I shan't be gone long. – You come too.

I'm going out to fetch the little calf
That's standing by the mother. It's so young
It totters when she licks it with her tongue.
I shan't be gone long. – You come too.

from A *Boy's Will*

Ghost House

I dwell in a lonely house I know
That vanished many a summer ago,
   And left no trace but the cellar walls,
   And a cellar in which the daylight falls
And the purple-stemmed wild raspberries grow.

O'er ruined fences the grapevines shield
The woods come back to the mowing field;
   The orchard tree has grown one copse
   Of new wood and old where the woodpecker chops;
10 The footpath down to the well is healed.

I dwell with a strangely aching heart
In that vanished abode there far apart
   On that disused and forgotten road
   That has no dust-bath now for the toad.
Night comes; the black bats tumble and dart;

The whippoorwill is coming to shout
And hush and cluck and flutter about:
   I hear him begin far enough away
   Full many a time to say his say
20 Before he arrives to say it out.

It is under the small, dim, summer star.
I know not who these mute folk are
   Who share the unlit place with me—
   Those stones out under the low-limbed tree
Doubtless bear names that the mosses mar.

They are tireless folk, but slow and sad—
Though two, close-keeping, are lass and lad—
   With none among them that ever sings,
   And yet, in view of how many things,
30 As sweet companions as might be had.

Waiting

Afield at dusk

What things for dream there are when specter-like,
Moving among tall haycocks lightly piled,
I enter alone upon the stubble field,
From which the laborers' voices late have died,
And in the antiphony of afterglow
And rising full moon, sit me down
Upon the full moon's side of the first haycock
And lose myself amid so many alike.

I dream upon the opposing lights of the hour,
10 Preventing shadow until the moon prevail;
I dream upon the nighthawks peopling heaven,
Each circling each with vague unearthly cry,
Or plunging headlong with fierce twang afar;
And on the bat's mute antics, who would seem
Dimly to have made out my secret place,
Only to lose it when he pirouettes,
And seek it endlessly with purblind haste;
On the last swallow's sweep; and on the rasp
In the abyss of odor and rustle at my back,
20 That, silenced by my advent, finds once more,
After an interval, his instrument,
And tries once – twice – and thrice if I be there;
And on the worn book of old-golden song
I brought not here to read, it seems, but hold
And freshen in this air of withering sweetness;
But on the memory of one absent, most,
For whom these lines when they shall greet her eye.

Mowing

There was never a sound beside the wood but one,
And that was my long scythe whispering to the ground.
What was it it whispered? I knew not well myself;
Perhaps it was something about the heat of the sun,
Something, perhaps, about the lack of sound—
And that was why it whispered and did not speak.
It was no dream of the gift of idle hours,
Or easy gold at the hand of fay or elf:
Anything more than the truth would have seemed too
weak
10 To the earnest love that laid the swale in rows,
Not without feeble-pointed spikes of flowers
(Pale orchises), and scared a bright green snake.
The fact is the sweetest dream that labor knows.
My long scythe whispered and left the hay to make.

The Tuft of Flowers

I went to turn the grass once after one
Who mowed it in the dew before the sun.

The dew was gone that made his blade so keen
Before I came to view the leveled scene.

I looked for him behind an isle of trees;
I listened for his whetstone on the breeze.

But he had gone his way, the grass all mown,
And I must be, as he had been – alone,

'As all must be,' I said within my heart,
10 'Whether they work together or apart.'

But as I said it, swift there passed me by
On noiseless wing a bewildered butterfly,

Seeking with memories grown dim o'er night
Some resting flower of yesterday's delight.

And once I marked his flight go round and round,
As where some flower lay withering on the ground.

And then he flew as far as eye could see,
And then on tremulous wing came back to me.

I thought of questions that have no reply,
20 And would have turned to toss the grass to dry;

But he turned first, and led my eye to look
At a tall tuft of flowers beside a brook,

A leaping tongue of bloom the scythe had spared
Beside a reedy brook the scythe had bared.

The mower in the dew had loved them thus,
By leaving them to flourish, not for us,

Nor yet to draw one thought of ours to him,
But from sheer morning gladness at the brim.

The butterfly and I had lit upon,
30 Nevertheless, a message from the dawn,

That made me hear the wakening birds around,
And hear his long scythe whispering to the ground,

And feel a spirit kindred to my own;
So that henceforth I worked no more alone;

But glad with him, I worked as with his aid,
And weary, sought at noon with him the shade;

And dreaming, as it were, held brotherly speech
With one whose thought I had not hoped to reach.

'Men work together,' I told him from the heart,
40 'Whether they work together or apart.'

In Hardwood Groves

The same leaves over and over again!
They fall from giving shade above,
To make one texture of faded brown
And fit the earth like a leather glove.

Before the leaves can mount again
To fill the trees with another shade,
They must go down past things coming up.
They must go down into the dark decayed.

They *must* be pierced by flowers and put
10 Beneath the feet of dancing flowers.
However it is in some other world
I know that this is the way in ours.

from *North of Boston*

Mending Wall

Something there is that doesn't love a wall,
That sends the frozen-ground-swell under it
And spills the upper boulders in the sun,
And makes gaps even two can pass abreast.
The work of hunters is another thing:
I have come after them and made repair
Where they have left not one stone on a stone,
But they would have the rabbit out of hiding,
To please the yelping dogs. The gaps I mean,
10 No one has seen them made or heard them made,
But at spring mending-time we find them there.
I let my neighbor know beyond the hill;
And on a day we meet to walk the line
And set the wall between us once again.
We keep the wall between us as we go.
To each the boulders that have fallen to each.
And some are loaves and some so nearly balls
We have to use a spell to make them balance:
'Stay where you are until our backs are turned!'
20 We wear our fingers rough with handling them.
Oh, just another kind of outdoor game,
One on a side. It comes to little more:
There where it is we do not need the wall:
He is all pine and I am apple orchard.
My apple trees will never get across
And eat the cones under his pines, I tell him.
He only says, 'Good fences make good neighbors.'
Spring is the mischief in me, and I wonder
If I could put a notion in his head:
30 '*Why* do they make good neighbors? Isn't it

Where there are cows? But here there are no cows.
Before I built a wall I'd ask to know
What I was walling in or walling out,
And to whom I was like to give offense.
Something there is that doesn't love a wall,
That wants it down.' I could say 'Elves' to him,
But it's not elves exactly, and I'd rather
He said it for himself. I see him there,
Bringing a stone grasped firmly by the top
40 In each hand, like an old-stone savage armed.
He moves in darkness as it seems to me,
Not of woods only and the shade of trees.
He will not go behind his father's saying,
And he likes having thought of it so well
He says again, 'Good fences make good neighbors.'

The Death of the Hired Man

Mary sat musing on the lamp-flame at the table,
Waiting for Warren. When she heard his step,
She ran on tiptoe down the darkened passage
To meet him in the doorway with the news
And put him on his guard. 'Silas is back.'
She pushed him outward with her through the door
And shut it after her. 'Be kind,' she said.
She took the market things from Warren's arms
And set them on the porch, then drew him down
10 To sit beside her on the wooden steps.

'When was I ever anything but kind to him?

But I'll not have the fellow back,' he said.
'I told him so last haying, didn't I?
If he left then, I said, that ended it.
What good is he? Who else will harbor him
At his age for the little he can do?
What help he is there's no depending on.
Off he goes always when I need him most.
He thinks he ought to earn a little pay,
20 Enough at least to buy tobacco with,
So he won't have to beg and be beholden.
"All right," I say, "I can't afford to pay
Any fixed wages, though I wish I could."
"Someone else can." "Then someone else will have to."
I shouldn't mind his bettering himself
If that was what it was. You can be certain,
When he begins like that, there's someone at him
Trying to coax him off with pocket money–
In haying time, when any help is scarce.
30 In winter he comes back to us. I'm done.'

'Sh! not so loud: he'll hear you,' Mary said.

'I want him to: he'll have to soon or late.'

'He's worn out. He's asleep beside the stove.
When I came up from Rowe's I found him here,
Huddled against the barn door fast asleep,
A miserable sight, and frightening, too–
You needn't smile – I didn't recognize him–
I wasn't looking for him – and he's changed.
Wait till you see.'

'Where did you say he'd been?'

40 'He didn't say. I dragged him to the house,

And gave him tea and tried to make him smoke.
I tried to make him talk about his travels.
Nothing would do: he just kept nodding off.'

'What did he say? Did he say anything?'

'But little.'

'Anything? Mary, confess
He said he'd come to ditch the meadow for me.'

'Warren!'

'But did he? I just want to know.'

'Of course he did. What would you have him say?
Surely you wouldn't grudge the poor old man
50 Some humble way to save his self-respect.
He added, if you really care to know,
He meant to clear the upper pasture, too.
That sounds like something you have heard before?
Warren, I wish you could have heard the way
He jumbled everything. I stopped to look
Two or three times – he made me feel so queer–
To see if he was talking in his sleep.
He ran on Harold Wilson – you remember–

The boy you had in haying four years since.
60 He's finished school, and teaching in his college.
Silas declares you'll have to get him back.
He says they two will make a team for work:
Between them they will lay this farm as smooth!
The way he mixed that in with other things.
He thinks young Wilson a likely lad, though daft
On education – you know how they fought
All through July under the blazing sun,
Silas up on the cart to build the load,
Harold along beside to pitch it on.'

70 'Yes, I took care to keep well out of earshot.'

'Well, those days trouble Silas like a dream.
You wouldn't think they would. How some things linger!
Harold's young college-boy's assurance piqued him.
After so many years he still keeps finding
Good arguments he sees he might have used.
I sympathize. I know just how it feels
To think of the right thing to say too late.
Harold's associated in his mind with Latin.
He asked me what I thought of Harold's saying
80 He studied Latin, like the violin,
Because he liked it – that an argument!
He said he couldn't make the boy believe
He could find water with a hazel prong–
Which showed how much good school had ever done
him.
He wanted to go over that. But most of all
He thinks if he could have another chance
To teach him how to build a load of hay–'

'I know, that's Silas' one accomplishment.
He bundles every forkful in its place,
90 And tags and numbers it for future reference,

So he can find and easily dislodge it
In the unloading. Silas does that well.
He takes it out in bunches like big birds' nests.
You never see him standing on the hay
He's trying to lift, straining to lift himself.'

'He thinks if he could teach him that, he'd be
Some good perhaps to someone in the world.
He hates to see a boy the fool of books.
Poor Silas, so concerned for other folk,
100 And nothing to look backward to with pride,
And nothing to look forward to with hope,
So now and never any different.'

Part of a moon was falling down the west,
Dragging the whole sky with it to the hills.
Its light poured softly in her lap. She saw it
And spread her apron to it. She put out her hand
Among the harplike morning-glory strings,
Taut with the dew from garden bed to eaves,
As if she played unheard some tenderness
110 That wrought on him beside her in the night.
'Warren,' she said, 'he has come home to die:
You needn't be afraid he'll leave you this time.'

'Home,' he mocked gently.

'Yes, what else but home?
It all depends on what you mean by home.
Of course he's nothing to us, any more
Than was the hound that came a stranger to us
Out of the woods, worn out upon the trail.'

'Home is the place where, when you have to go there,
They have to take you in.'

'I should have called it
120 Something you somehow haven't to deserve.'

Warren leaned out and took a step or two,
Picked up a little stick, and brought it back
And broke it in his hand and tossed it by.
'Silas has better claim on us you think
Than on his brother? Thirteen little miles
As the road winds would bring him to his door.
Silas has walked that far no doubt today.

Why doesn't he go there? His brother's rich,
A somebody – director in the bank.'

130 'He never told us that.'

'We know it, though.'

'I think his brother ought to help, of course.
I'll see to that if there is need. He ought of right
To take him in, and might be willing to–
He may be better than appearances.
But have some pity on Silas. Do you think
If he had any pride in claiming kin
Or anything he looked for from his brother,
He'd keep so still about him all this time?'

'I wonder what's between them.'

'I can tell you.
140 Silas is what he is – we wouldn't mind him–
But just the kind that kinsfolk can't abide.
He never did a thing so very bad.
He don't know why he isn't quite as good
As anybody. Worthless though he is,
He won't be made ashamed to please his brother.'

'*I* can't think Si ever hurt anyone.'

'No, but he hurt my heart the way he lay
And rolled his old head on that sharp-edged chair-back.
He wouldn't let me put him on the lounge.

150 You must go in and see what you can do.
I made the bed up for him there tonight.
You'll be surprised at him – how much he's broken.
His working days are done; I'm sure of it.'

'I'd not be in a hurry to say that.'

'I haven't been. Go, look, see for yourself.
But, Warren, please remember how it is:
He's come to help you ditch the meadow.
He has a plan. You mustn't laugh at him.
He may not speak of it, and then he may.
160 I'll sit and see if that small sailing cloud
Will hit or miss the moon.'

It hit the moon.
Then there were three there, making a dim row,
The moon, the little silver cloud, and she.

Warren returned – too soon, it seemed to her—
Slipped to her side, caught up her hand and waited.

'Warren?' she questioned.

'Dead,' was all he answered.

Home Burial

He saw her from the bottom of the stairs
Before she saw him. She was starting down,
Looking back over her shoulder at some fear.
She took a doubtful step and then undid it
To raise herself and look again. He spoke
Advancing toward her: 'What is it you see
From up there always? – for I want to know.'
She turned and sank upon her skirts at that,
And her face changed from terrified to dull.
10 He said to gain time: 'What is it you see?'
Mounting until she cowered under him.
'I will find out now – you must tell me, dear.'
She, in her place, refused him any help,
With the least stiffening of her neck and silence.
She let him look, sure that he wouldn't see,
Blind creature; and awhile he didn't see.
But at last he murmured, 'Oh,' and again, 'Oh.'

'What is it – what?' she said.

'Just that I see.'

'You don't,' she challenged. 'Tell me what it is.'

20 'The wonder is I didn't see at once.
I never noticed it from here before.
I must be wonted to it – that's the reason.
The little graveyard where my people are!
So small the window frames the whole of it.
Not so much larger than a bedroom, is it?
There are three stones of slate and one of marble,
Broad-shouldered little slabs there in the sunlight
On the sidehill. We haven't to mind *those.*
But I understand: it is not the stones,

30 But the child's mound–'

'Don't, don't, don't,
don't,' she cried.

She withdrew, shrinking from beneath his arm
That rested on the banister, and slid downstairs;
And turned on him with such a daunting look,
He said twice over before he knew himself:
'Can't a man speak of his own child he's lost?'

'Not you – Oh, where's my hat? Oh, I don't need it!
I must get out of here. I must get air.–
I don't know rightly whether any man can.'

'Amy! Don't go to someone else this time.
40 Listen to me. I won't come down the stairs.'
He sat and fixed his chin between his fists.
'There's something I should like to ask you, dear.'

'You don't know how to ask it.'

'Help me, then.'

Her fingers moved the latch for all reply.

'My words are nearly always an offense.
I don't know how to speak of anything
So as to please you. But I might be taught,
I should suppose. I can't say I see how.
A man must partly give up being a man
50 With womenfolk. We could have some arrangement
By which I'd bind myself to keep hands off
Anything special you're a-mind to name.
Though I don't like such things 'twixt those that love.
Two that don't love can't live together without them.
But two that do can't live together with them.'
She moved the latch a little. 'Don't – don't go.
Don't carry it to someone else this time.

Tell me about it if it's something human.
Let me into your grief. I'm not so much
60 Unlike other folks as your standing there
Apart would make me out. Give me my chance.
I do think, though, you overdo it a little.
What was it brought you up to think it the thing
To take your mother-loss of a first child
So inconsolably – in the face of love.
You'd think his memory might be satisfied—'

'There you go sneering now!'

'I'm not, I'm not!
You make me angry. I'll come down to you.
God, what a woman! And it's come to this,
70 A man can't speak of his own child that's dead.'

'You can't because you don't know how to speak.
If you had any feelings, you that dug
With your own hand – how could you? – his little grave;
I saw you from that very window there,
Making the gravel leap and leap in air,
Leap up, like that, like that, and land so lightly
And roll back down the mound beside the hole.
I thought, Who is that man? I didn't know you.
And I crept down the stairs and up the stairs
80 To look again, and still your spade kept lifting.
Then you came in. I heard your rumbling voice

Out in the kitchen, and I don't know why,
But I went near to see with my own eyes.
You could sit there with the stains on your shoes
Of the fresh earth from your own baby's grave
And talk about your everyday concerns.
You had stood the spade up against the wall
Outside there in the entry, for I saw it.'

'I shall laugh the worst laugh I ever laughed.
90 I'm cursed. God, if I don't believe I'm cursed.'

'I can repeat the very words you were saying:
"Three foggy mornings and one rainy day
Will rot the best birch fence a man can build."
Think of it, talk like that at such a time!
What had how long it takes a birch to rot
To do with what was in the darkened parlor?
You *couldn't* care! The nearest friends can go
With anyone to death, comes so far short
They might as well not try to go at all.
100 No, from the time when one is sick to death,
One is alone, and he dies more alone.
Friends make pretense of following to the grave,
But before one is in it, their minds are turned
And making the best of their way back to life
And living people, and things they understand.
But the world's evil. I won't have grief so
If I can change it. Oh, I won't, I won't!'

'There, you have said it all and you feel better.
You won't go now. You're crying. Close the door.
110 The heart's gone out of it: why keep it up?
Amy! There's someone coming down the road!'

'*You* – oh, you think the talk is all. I must go—
Somewhere out of this house. How can I make you—'

'If – you – do!' She was opening the door wider.
'Where do you mean to go? First tell me that.
I'll follow and bring you back by force. I *will*!—'

The Black Cottage

We chanced in passing by that afternoon
To catch it in a sort of special picture
Among tar-banded ancient cherry trees,
Set well back from the road in rank lodged grass,
The little cottage we were speaking of,
A front with just a door between two windows,
Fresh painted by the shower a velvet black.
We paused, the minister and I, to look.
He made as if to hold it at arm's length
10 Or put the leaves aside that framed it in.
'Pretty,' he said. 'Come in. No one will care.'
The path was a vague parting in the grass
That led us to a weathered windowsill.
We pressed our faces to the pane. 'You see,' he said,
'Everything's as she left it when she died.
Her sons won't sell the house or the things in it.
They say they mean to come and summer here
Where they were boys. They haven't come this year.
They live so far away – one is out West–
20 It will be hard for them to keep their word.
Anyway they won't have the place disturbed.'
A buttoned haircloth lounge spread scrolling arms
Under a crayon portrait on the wall,
Done sadly from an old daguerreotype.
'That was the father as he went to war.
She always, when she talked about the war,
Sooner or later came and leaned, half knelt,
Against the lounge beside it, though I doubt
If such unlifelike lines kept power to stir
30 Anything in her after all the years.
He fell at Gettysburg or Fredericksburg,
I ought to know – it makes a difference which:
Fredericksburg wasn't Gettysburg, of course.

But what I'm getting to is how forsaken
A little cottage this has always seemed;
Since she went, more than ever, but before–
I don't mean altogether by the lives
That had gone out of it, the father first,
Then the two sons, till she was left alone.
40 (Nothing could draw her after those two sons.
She valued the considerate neglect
She had at some cost taught them after years.)
I mean by the world's having passed it by–
As we almost got by this afternoon.
It always seems to me a sort of mark
To measure how far fifty years have brought us.
Why not sit down if you are in no haste?
These doorsteps seldom have a visitor.
The warping boards pull out their own old nails
50 With none to tread and put them in their place.
She had her own idea of things, the old lady.
And she liked talk. She had seen Garrison
And Whittier, and had her story of them.
One wasn't long in learning that she thought,
Whatever else the Civil War was for,
It wasn't just to keep the States together,
Nor just to free the slaves, though it did both.
She wouldn't have believed those ends enough
To have given outright for them all she gave.
60 Her giving somehow touched the principle
That all men are created free and equal.
And to hear her quaint phrases – so removed
From the world's view today of all those things.
That's a hard mystery of Jefferson's.
What did he mean? Of course the easy way
Is to decide it simply isn't true.
It may not be. I heard a fellow say so.
But never mind, the Welshman got it planted

Where it will trouble us a thousand years.
70 Each age will have to reconsider it.
You couldn't tell her what the West was saying,
And what the South, to her serene belief.
She had some art of hearing and yet not
Hearing the latter wisdom of the world.
White was the only race she ever knew.
Black she had scarcely seen, and yellow never.
But how could they be made so very unlike
By the same hand working in the same stuff?
She had supposed the war decided that.
80 What are you going to do with such a person?
Strange how such innocence gets its own way.
I shouldn't be surprised if in this world
It were the force that would at last prevail.
Do you know but for her there was a time
When, to please younger members of the church,
Or rather say non-members in the church,
Whom we all have to think of nowadays,
I would have changed the Creed a very little?
Not that she ever had to ask me not to;
90 It never got so far as that; but the bare thought
Of her old tremulous bonnet in the pew,
And of her half asleep, was too much for me.
Why, I might wake her up and startle her.
It was the words "descended into Hades"
That seemed too pagan to our liberal youth.
You know they suffered from a general onslaught.
And well, if they weren't true why keep right on
Saying them like the heathen? We could drop them.
Only – there was the bonnet in the pew.
100 Such a phrase couldn't have meant much to her.
But suppose she had missed it from the Creed,
As a child misses the unsaid Good-night
And falls asleep with heartache – how should *I* feel?

I'm just as glad she made me keep hands off,
For, dear me, why abandon a belief
Merely because it ceases to be true.
Cling to it long enough, and not a doubt
It will turn true again, for so it goes.
Most of the change we think we see in life
110 Is due to truths being in and out of favor.
As I sit here, and oftentimes, I wish
I could be monarch of a desert land
I could devote and dedicate forever
To the truths we keep coming back and back to.
So desert it would have to be, so walled
By mountain ranges half in summer snow,
No one would covet it or think it worth
The pains of conquering to force change on.
Scattered oases where men dwelt, but mostly
120 Sand dunes held loosely in tamarisk
Blown over and over themselves in idleness.
Sand grains should sugar in the natal dew
The babe born to the desert, the sandstorm
Retard mid-waste my cowering caravans–

'There are bees in this wall.' He struck the clapboards,
Fierce heads looked out; small bodies pivoted.
We rose to go. Sunset blazed on the windows.

After Apple-Picking

My long two-pointed ladder's sticking through a tree
Toward heaven still,
And there's a barrel that I didn't fill
Beside it, and there may be two or three
Apples I didn't pick upon some bough.
But I am done with apple-picking now.
Essence of winter sleep is on the night,
The scent of apples: I am drowsing off.
I cannot rub the strangeness from my sight
10 I got from looking through a pane of glass
I skimmed this morning from the drinking trough
And held against the world of hoary grass.
It melted, and I let it fall and break.
But I was well
Upon my way to sleep before it fell,
And I could tell
What form my dreaming was about to take.
Magnified apples appear and disappear,
Stem end and blossom end,
20 And every fleck of russet showing clear.
My instep arch not only keeps the ache,
It keeps the pressure of a ladder-round.
I feel the ladder sway as the boughs bend.
And I keep hearing from the cellar bin
The rumbling sound
Of load on load of apples coming in.
For I have had too much
Of apple-picking: I am overtired
Of the great harvest I myself desired.
30 There were ten thousand thousand fruit to touch,
Cherish in hand, lift down, and not let fall.
For all
That struck the earth,

No matter if not bruised or spiked with stubble,
Went surely to the cider-apple heap
As of no worth.
One can see what will trouble
This sleep of mine, whatever sleep it is.
Were he not gone,
40 The woodchuck could say whether it's like his
Long sleep, as I describe its coming on,
Or just some human sleep.

The Wood-Pile

Out walking in the frozen swamp one gray day,
I paused and said, 'I will turn back from here.
No, I will go on farther – and we shall see.'
The hard snow held me, save where now and then
One foot went through. The view was all in lines
Straight up and down of tall slim trees
Too much alike to mark or name a place by
So as to say for certain I was here
Or somewhere else: I was just far from home.
10 A small bird flew before me. He was careful
To put a tree between us when he lighted,
And say no word to tell me who he was
Who was so foolish as to think what *he* thought.
He thought that I was after him for a feather–
The white one in his tail; like one who takes
Everything said as personal to himself.
One flight out sideways would have undeceived him.
And then there was a pile of wood for which
I forgot him and let his little fear
20 Carry him off the way I might have gone,
Without so much as wishing him good-night.
He went behind it to make his last stand.

It was a cord of maple, cut and split
And piled – and measured, four by four by eight.
And not another like it could I see.
No runner tracks in this year's snow looped near it.
And it was older sure than this year's cutting,
Or even last year's or the year's before.
The wood was gray and the bark warping off it
30 And the pile somewhat sunken. Clematis
Had wound strings round and round it like a bundle.
What held it, though, on one side was a tree
Still growing, and on one a stake and prop,
These latter about to fall. I thought that only
Someone who lived in turning to fresh tasks
Could so forget his handiwork on which
He spent himself, the labor of his ax,
And leave it there far from a useful fireplace
To warm the frozen swamp as best it could
40 With the slow smokeless burning of decay.

from *Mountain Interval*

The Road Not Taken

Two roads diverged in a yellow wood,
And sorry I could not travel both
And be one traveler, long I stood
And looked down one as far as I could
To where it bent in the undergrowth;

Then took the other, as just as fair,
And having perhaps the better claim,
Because it was grassy and wanted wear;
Though as for that, the passing there
10 Had worn them really about the same,

And both that morning equally lay
In leaves no step had trodden black.
Oh, I kept the first for another day!
Yet knowing how way leads on to way,
I doubted if I should ever come back.

I shall be telling this with a sigh
Somewhere ages and ages hence:
Two roads diverged in a wood, and I–
I took the one less traveled by,
20 And that has made all the difference.

The Oven Bird

There is a singer everyone has heard,
Loud, a mid-summer and a mid-wood bird,
Who makes the solid tree trunks sound again.
He says that leaves are old and that for flowers
Mid-summer is to spring as one to ten.
He says the early petal-fall is past,
When pear and cherry bloom went down in showers
On sunny days a moment overcast;
And comes that other fall we name the fall.
10 He says the highway dust is over all.
The bird would cease and be as other birds
But that he knows in singing not to sing.
The question that he frames in all but words
Is what to make of a diminished thing.

CORK CITY LIBRARY

Birches

When I see birches bend to left and right
Across the lines of straighter darker trees,
I like to think some boy's been swinging them.
But swinging doesn't bend them down to stay
As ice storms do. Often you must have seen them
Loaded with ice a sunny winter morning
After a rain. They click upon themselves
As the breeze rises, and turn many-colored
As the stir cracks and crazes their enamel.
10 Soon the sun's warmth makes them shed crystal shells
Shattering and avalanching on the snow crust—
Such heaps of broken glass to sweep away
You'd think the inner dome of heaven had fallen.
They are dragged to the withered bracken by the load,
And they seem not to break; though once they are bowed
So low for long, they never right themselves:
You may see their trunks arching in the woods
Years afterwards, trailing their leaves on the ground
Like girls on hands and knees that throw their hair
20 Before them over their heads to dry in the sun.
But I was going to say when Truth broke in
With all her matter of fact about the ice storm,
I should prefer to have some boy bend them
As he went out and in to fetch the cows—
Some boy too far from town to learn baseball,
Whose only play was what he found himself,
Summer or winter, and could play alone.
One by one he subdued his father's trees
By riding them down over and over again
30 Until he took the stiffness out of them,
And not one but hung limp, not one was left

For him to conquer. He learned all there was
To learn about not launching out too soon
And so not carrying the tree away
Clear to the ground. He always kept his poise
To the top branches, climbing carefully
With the same pains you use to fill a cup
Up to the brim, and even above the brim.
Then he flung outward, feet first, with a swish,
40 Kicking his way down through the air to the ground.
So was I once myself a swinger of birches.
And so I dream of going back to be.
It's when I'm weary of considerations,
And life is too much like a pathless wood
Where your face burns and tickles with the cobwebs
Broken across it, and one eye is weeping
From a twig's having lashed across it open.
I'd like to get away from earth awhile
And then come back to it and begin over.
50 May no fate willfully misunderstand me
And half grant what I wish and snatch me away
Not to return. Earth's the right place for love:
I don't know where it's likely to go better.
I'd like to go by climbing a birch tree,
And climb black branches up a snow-white trunk
Toward heaven, till the tree could bear no more,
But dipped its top and set me down again.
That would be good both going and coming back.
One could do worse than be a swinger of birches.

The Cow in Apple Time

Something inspires the only cow of late
To make no more of a wall than an open gate,
And think no more of wall-builders than fools.
Her face is flecked with pomace and she drools
A cider syrup. Having tasted fruit,
She scorns a pasture withering to the root.
She runs from tree to tree where lie and sweeten
The windfalls spiked with stubble and worm-eaten.
She leaves them bitten when she has to fly.
10 She bellows on a knoll against the sky.
Her udder shrivels and the milk goes dry.

An Encounter

Once on the kind of day called 'weather breeder,'
When the heat slowly hazes and the sun
By its own power seems to be undone,
I was half boring through, half climbing through
A swamp of cedar. Choked with oil of cedar
And scurf of plants, and weary and overheated,
And sorry I ever left the road I knew,
I paused and rested on a sort of hook
That had me by the coat as good as seated,
10 And since there was no other way to look,
Looked up toward heaven, and there against the blue,
Stood over me a resurrected tree,
A tree that had been down and raised again–
A barkless specter. He had halted too,
As if for fear of treading upon me.
I saw the strange position of his hands–
Up at his shoulders, dragging yellow strands
Of wire with something in it from men to men.
'You here?' I said. 'Where aren't you nowadays?
20 And what's the news you carry – if you know?
And tell me where you're off for – Montreal?
Me? I'm not off for anywhere at all.
Sometimes I wander out of beaten ways
Half looking for the orchid Calypso.'

'Out, Out—'

The buzz saw snarled and rattled in the yard
And made dust and dropped stove-length sticks of wood,
Sweet-scented stuff when the breeze drew across it.
And from there those that lifted eyes could count
Five mountain ranges one behind the other
Under the sunset far into Vermont.
And the saw snarled and rattled, snarled and rattled,
As it ran light, or had to bear a load.
And nothing happened: day was all but done.
10 Call it a day, I wish they might have said
To please the boy by giving him the half hour
That a boy counts so much when saved from work.
His sister stood beside them in her apron
To tell them 'Supper.' At the word, the saw,
As if to prove saws knew what supper meant,
Leaped out at the boy's hand, or seemed to leap—
He must have given the hand. However it was,
Neither refused the meeting. But the hand!
The boy's first outcry was a rueful laugh,
20 As he swung toward them holding up the hand,
Half in appeal, but half as if to keep
The life from spilling. Then the boy saw all—
Since he was old enough to know, big boy
Doing a man's work, though a child at heart—
He saw all spoiled. 'Don't let him cut my hand off—
The doctor, when he comes. Don't let him, sister!'
So. But the hand was gone already.
The doctor put him in the dark of ether.
He lay and puffed his lips out with his breath.
30 And then – the watcher at his pulse took fright.
No one believed. They listened at his heart.

Little – less – nothing! – and that ended it.
No more to build on there. And they, since they
Were not the one dead, turned to their affairs.

The Sound of Trees

I wonder about the trees.
Why do we wish to bear
Forever the noise of these
More than another noise
So close to our dwelling place?
We suffer them by the day
Till we lose all measure of pace,
And fixity in our joys,
And acquire a listening air.
10 They are that that talks of going
But never gets away;
And that talks no less for knowing,
As it grows wiser and older,
That now it means to stay.
My feet tug at the floor
And my head sways to my shoulder
Sometimes when I watch trees sway,
From the window or the door.
I shall set forth for somewhere,
20 I shall make the reckless choice
Some day when they are in voice
And tossing so as to scare
The white clouds over them on.
I shall have less to say,
But I shall be gone.

from *New Hampshire*

The Ax-Helve

I've known ere now an interfering branch
Of alder catch my lifted ax behind me.
But that was in the woods, to hold my hand
From striking at another alder's roots,
And that was, as I say, an alder branch.
This was a man, Baptiste, who stole one day
Behind me on the snow in my own yard
Where I was working at the chopping block,
And cutting nothing not cut down already.
10 He caught my ax expertly on the rise,
When all my strength put forth was in his favor,
Held it a moment where it was, to calm me,
Then took it from me – and I let him take it.
I didn't know him well enough to know
What it was all about. There might be something
He had in mind to say to a bad neighbor
He might prefer to say to him disarmed.
But all he had to tell me in French-English
Was what he thought of – not me, but my ax,
20 Me only as I took my ax to heart.
It was the bad ax-helve someone had sold me–
'Made on machine,' he said, plowing the grain
With a thick thumbnail to show how it ran
Across the handle's long-drawn serpentine,
Like the two strokes across a dollar sign.
'You give her one good crack, she's snap raght off.
Den where's your hax-ead flying t'rough de hair?'
Admitted; and yet, what was that to him?

'Come on my house and I put you one in
30 What's las' awhile – good hick'ry what's grow crooked,
De second growt' I cut myself – tough, tough!'

Something to sell? That wasn't how it sounded.

'Den when you say you come? It's cost you nothing.
Tonaght?'

As well tonight as any night.

Beyond an over-warmth of kitchen stove
My welcome differed from no other welcome.
Baptiste knew best why I was where I was.
So long as he would leave enough unsaid,
I shouldn't mind his being overjoyed
40 (If overjoyed he was) at having got me
Where I must judge if what he knew about an ax
That not everybody else knew was to count
For nothing in the measure of a neighbor.
Hard if, though cast away for life with Yankees,
A Frenchman couldn't get his human rating!

Mrs Baptiste came in and rocked a chair
That had as many motions as the world:
One back and forward, in and out of shadow,
That got her nowhere; one more gradual,
50 Sideways, that would have run her on the stove
In time, had she not realized her danger
And caught herself up bodily, chair and all,
And set herself back where she started from.
'She ain't spick too much Henglish – dat's too bad.'
I was afraid, in brightening first on me,
Then on Baptiste, as if she understood
What passed between us, she was only feigning.

Baptiste was anxious for her; but no more
Than for himself, so placed he couldn't hope
60 To keep his bargain of the morning with me
In time to keep me from suspecting him
Of really never having meant to keep it.

Needlessly soon he had his ax-helves out,
A quiverful to choose from, since he wished me
To have the best he had, or had to spare—
Not for me to ask which, when what he took
Had beauties he had to point me out at length
To insure their not being wasted on me.
He liked to have it slender as a whipstock,
70 Free from the least knot, equal to the strain
Of bending like a sword across the knee.
He showed me that the lines of a good helve
Were native to the grain before the knife
Expressed them, and its curves were no false curves
Put on it from without. And there its strength lay
For the hard work. He chafed its long white body
From end to end with his rough hand shut round it.
He tried it at the eyehole in the ax-head.
'Hahn, hahn,' he mused, 'don't need much taking down.'
80 Baptiste knew how to make a short job long
For love of it, and yet not waste time either.

Do you know, what we talked about was knowledge?
Baptiste on his defense about the children
He kept from school, or did his best to keep—
Whatever school and children and our doubts
Of laid-on education had to do
With the curves of his ax-helves and his having
Used these unscrupulously to bring me
To see for once the inside of his house.
90 Was I desired in friendship, partly as someone
To leave it to, whether the right to hold

Such doubts of education should depend
Upon the education of those who held them?

But now he brushed the shavings from his knee
And stood the ax there on its horse's hoof,
Erect, but not without its waves, as when
The snake stood up for evil in the Garden–
Top-heavy with a heaviness his short,
Thick hand made light of, steel-blue chin drawn down
100 And in a little – a French touch in that.
Baptiste drew back and squinted at it, pleased:
'See how she's cock her head!'

To E.T.

I slumbered with your poems on my breast,
Spread open as I dropped them half-read through
Like dove wings on a figure on a tomb,
To see if in a dream they brought of you

I might not have the chance I missed in life
Through some delay, and call you to your face
First soldier, and then poet, and then both,
Who died a soldier-poet of your race.

I meant, you meant, that nothing should remain
10 Unsaid between us, brother, and this remained–
And one thing more that was not then to say:
The Victory for what it lost and gained.

You went to meet the shell's embrace of fire
On Vimy Ridge; and when you fell that day
The war seemed over more for you than me,
But now for me than you – the other way.

How over, though, for even me who knew
The foe thrust back unsafe beyond the Rhine,
If I was not to speak of it to you
20 And see you pleased once more with words of mine?

Stopping by Woods on a Snowy Evening

Whose woods these are I think I know.
His house is in the village, though;
He will not see me stopping here
To watch his woods fill up with snow.

My little horse must think it queer
To stop without a farmhouse near
Between the woods and frozen lake
The darkest evening of the year.

He gives his harness bells a shake
10 To ask if there is some mistake.
The only other sound's the sweep
Of easy wind and downy flake.

The woods are lovely, dark, and deep,
But I have promises to keep,
And miles to go before I sleep,
And miles to go before I sleep.

Two Look at Two

Love and forgetting might have carried them
A little further up the mountainside
With night so near, but not much further up.
They must have halted soon in any case
With thoughts of the path back, how rough it was
With rock and washout, and unsafe in darkness;
When they were halted by a tumbled wall
With barbed-wire binding. They stood facing this,
Spending what onward impulse they still had
10 In one last look the way they must not go,
On up the failing path, where, if a stone
Or earthslide moved at night, it moved itself;
No footstep moved it.' This is all,' they sighed,
'Good-night to woods.' But not so; there was more.
A doe from round a spruce stood looking at them
Across the wall, as near the wall as they.
She saw them in their field, they her in hers.
The difficulty of seeing what stood still,
Like some up-ended boulder split in two,
20 Was in her clouded eyes: they saw no fear there.
She seemed to think that, two thus, they were safe.
Then, as if they were something that, though strange,
She could not trouble her mind with too long,
She sighed and passed unscared along the wall.
'*This*, then, is all. What more is there to ask?'
But no, not yet. A snort to bid them wait.
A buck from round the spruce stood looking at them
Across the wall, as near the wall as they.
This was an antlered buck of lusty nostril,
30 Not the same doe come back into her place.
He viewed them quizzically with jerks of head,
As if to ask, 'Why don't you make some motion?
Or give some sign of life? Because you can't.

I doubt if you're as living as you look.'
Thus till he had them almost feeling dared
To stretch a proffering hand – and a spell-breaking.
Then he too passed unscared along the wall.
Two had seen two, whichever side you spoke from.
'This *must* be all.' It was all. Still they stood,
40 A great wave from it going over them,
As if the earth in one unlooked-for favor
Had made them certain earth returned their love.

Gathering Leaves

Spades take up leaves
No better than spoons,
And bags full of leaves
Are light as balloons.

I make a great noise
Of rustling all day
Like rabbit and deer
Running away.

But the mountains I raise
10 Elude my embrace,
Flowing over my arms
And into my face.

I may load and unload
Again and again
Till I fill the whole shed,
And what have I then?

Next to nothing for weight;
And since they grew duller
From contact with earth,
20 Next to nothing for color.

Next to nothing for use.
But a crop is a crop,
And who's to say where
The harvest shall stop?

from *West-Running Brook*

Tree at My Window

Tree at my window, window tree,
My sash is lowered when night comes on;
But let there never be curtain drawn
Between you and me.

Vague dream-head lifted out of the ground,
And thing next most diffuse to cloud,
Not all your light tongues talking aloud
Could be profound.

But, tree, I have seen you taken and tossed,
10 And if you have seen me when I slept,
You have seen me when I was taken and swept
And all but lost.

That day she put our heads together,
Fate had her imagination about her,
Your head so much concerned with outer,
Mine with inner, weather.

Acquainted with the Night

I have been one acquainted with the night.
I have walked out in rain – and back in rain.
I have outwalked the furthest city light.

I have looked down the saddest city lane.
I have passed by the watchman on his beat
And dropped my eyes, unwilling to explain.

I have stood still and stopped the sound of feet
When far away an interrupted cry
Came over houses from another street,

10 But not to call me back or say good-by;
And further still at an unearthly height
One luminary clock against the sky

Proclaimed the time was neither wrong nor right.
I have been one acquainted with the night.

A Soldier

He is that fallen lance that lies as hurled,
That lies unlifted now, come dew, come rust,
But still lies pointed as it plowed the dust.
If we who sight along it round the world,
See nothing worthy to have been its mark,
It is because like men we look too near,
Forgetting that as fitted to the sphere,
Our missiles always make too short an arc.
They fall, they rip the grass, they intersect
10 The curve of earth, and striking, break their own;
They make us cringe for metal-point on stone.
But this we know, the obstacle that checked
And tripped the body, shot the spirit on
Further than target ever showed or shone.

from A *Further Range*

Desert Places

Snow falling and night falling fast, oh, fast
In a field I looked into going past,
And the ground almost covered smooth in snow,
But a few weeds and stubble showing last.

The woods around it have it – it is theirs.
All animals are smothered in their lairs.
I am too absent-spirited to count;
The loneliness includes me unawares.

And lonely as it is, that loneliness
10 Will be more lonely ere it will be less—
A blanker whiteness of benighted snow
With no expression, nothing to express.

They cannot scare me with their empty spaces
Between stars – on stars where no human race is.
I have it in me so much nearer home
To scare myself with my own desert places.

A Leaf-Treader

I have been treading on leaves all day until I am autumn-
    tired.
God knows all the color and form of leaves I have trodden
    on and mired.
Perhaps I have put forth too much strength and been too
    fierce from fear.
I have safely trodden underfoot the leaves of another year.

All summer long they were overhead, more lifted up
    than I.
To come to their final place in earth they had to pass
    me by.
All summer long I thought I heard them threatening
    under their breath.
And when they came it seemed with a will to carry me
    with them to death.

They spoke to the fugitive in my heart as if it were leaf to
    leaf.
10 They tapped at my eyelids and touched my lips with an
    invitation to grief.
But it was no reason I had to go because they had to go.
Now up, my knee, to keep on top of another year of snow.

Neither Out Far nor In Deep

The people along the sand
All turn and look one way.
They turn their back on the land.
They look at the sea all day.

As long as it takes to pass
A ship keeps raising its hull;
The wetter ground like glass
Reflects a standing gull.

The land may vary more;
10 But wherever the truth may be–
The water comes ashore,
And the people look at the sea.

They cannot look out far.
They cannot look in deep.
But when was that ever a bar
To any watch they keep?

There Are Roughly Zones

We sit indoors and talk of the cold outside.
And every gust that gathers strength and heaves
Is a threat to the house. But the house has long been tried.
We think of the tree. If it never again has leaves,
We'll know, we say, that this was the night it died.
It is very far north, we admit, to have brought the peach.
What comes over a man, is it soul or mind–
That to no limits and bounds he can stay confined?
You would say his ambition was to extend the reach
10 Clear to the Arctic of every living kind.
Why is his nature forever so hard to teach
That though there is no fixed line between wrong and
right,
There are roughly zones whose laws must be obeyed?
There is nothing much we can do for the tree tonight,
But we can't help feeling more than a little betrayed
That the northwest wind should rise to such a height
Just when the cold went down so many below.
The tree has no leaves and may never have them again.
We must wait till some months hence in the spring to
know.
20 But if it is destined never again to grow,
It can blame this limitless trait in the hearts of men.

from A *Witness Tree*

The Most of It

He thought he kept the universe alone;
For all the voice in answer he could wake
Was but the mocking echo of his own
From some tree-hidden cliff across the lake.
Some morning from the boulder-broken beach
He would cry out on life, that what it wants
Is not its own love back in copy speech,
But counter-love, original response.
And nothing ever came of what he cried
10 Unless it was the embodiment that crashed
In the cliff's talus on the other side,
And then in the far-distant water splashed,
But after a time allowed for it to swim,
Instead of proving human when it neared
And someone else additional to him,
As a great buck it powerfully appeared,
Pushing the crumpled water up ahead,
And landed pouring like a waterfall,
And stumbled through the rocks with horny tread,
20 And forced the underbrush – and that was all.

A Considerable Speck
(Microscopic)

A speck that would have been beneath my sight
On any but a paper sheet so white
Set off across what I had written there.
And I had idly poised my pen in air
To stop it with a period of ink,
When something strange about it made me think.
This was no dust speck by my breathing blown,
But unmistakably a living mite
With inclinations it could call its own.
10 It paused as with suspicion of my pen,
And then came racing wildly on again
To where my manuscript was not yet dry;
Then paused again and either drank or smelt–
With loathing, for again it turned to fly.
Plainly with an intelligence I dealt.
It seemed too tiny to have room for feet,
Yet must have had a set of them complete
To express how much it didn't want to die.
It ran with terror and with cunning crept.
20 It faltered: I could see it hesitate;
Then in the middle of the open sheet
Cower down in desperation to accept
Whatever I accorded it of fate.
I have none of the tenderer-than-thou
Collectivistic regimenting love

With which the modern world is being swept.
But this poor microscopic item now!
Since it was nothing I knew evil of
I let it lie there till I hope it slept.
30 I have a mind myself and recognize
Mind when I meet with it in any guise.
No one can know how glad I am to find
On any sheet the least display of mind.

from *Steeple Bush*

A Young Birch

The birch begins to crack its outer sheath
Of baby green and show the white beneath,
As whosoever likes the young and slight
May well have noticed. Soon entirely white
To double day and cut in half the dark
It will stand forth, entirely white in bark,
And nothing but the top a leafy green–
The only native tree that dares to lean,
Relying on its beauty, to the air.
10 (Less brave perhaps than trusting are the fair.)
And someone reminiscent will recall
How once in cutting brush along the wall
He spared it from the number of the slain,
At first to be no bigger than a cane,
And then no bigger than a fishing pole,
But now at last so obvious a bole
The most efficient help you ever hired
Would know that it was there to be admired,
And zeal would not be thanked that cut it down
20 When you were reading books or out of town.
It was a thing of beauty and was sent
To live its life out as an ornament.

An Unstamped Letter in Our Rural Letter Box

Last night your watchdog barked all night,
So once you rose and lit the light.
It wasn't someone at your locks.
No, in your rural letter box
I leave this note without a stamp
To tell you it was just a tramp
Who used your pasture for a camp.
There, pointed like the pip of spades,
The young spruce made a suite of glades
10 So regular that in the dark
The place was like a city park.
There I elected to demur
Beneath a low-slung juniper
That like a blanket to my chin
Kept some dew out and some heat in,
Yet left me freely face to face
All night with universal space.
It may have been at two o'clock
That under me a point of rock
20 Developed in the grass and fern,
And as I woke afraid to turn
Or so much as uncross my feet,
Lest having wasted precious heat
I never should again be warmed,
The largest firedrop ever formed
From two stars having coalesced
Went streaking molten down the west.
And then your tramp astrologer
From seeing this undoubted stir
30 In Heaven's firm-set firmament,
Himself had the equivalent,
Only within. Inside the brain

Two memories that long had lain
Now quivered toward each other, lipped
Together, and together slipped;
And for a moment all was plain
That men have thought about in vain.
Please, my involuntary host,
Forgive me if I seem to boast.
40 'Tis possible you may have seen,
Albeit through a rusty screen,
The same sign Heaven showed your guest.
Each knows his own discernment best.
You have had your advantages.
Things must have happened to you, yes,
And have occurred to you no doubt,
If not indeed from sleeping out,
Then from the work you went about
In farming well - or pretty well.
50 And it is partly to compel
Myself, *in forma pauperis*,
To say as much I write you this.

To an Ancient

Your claims to immortality were two.
The one you made, the other one you grew.
Sorry to have no name for you but You.

We never knew exactly where to look,
But found one in the delta of a brook,
One in a cavern where you used to cook.

Coming on such an ancient human trace
Seems as expressive of the human race
As meeting someone living, face to face.

10 We date you by your depth in silt and dust
Your probable brute nature is discussed.
At which point we are totally nonplussed.

You made the eolith, you grew the bone,
The second more peculiarly your own,
And likely to have been enough alone.

You make me ask if I would go to time
Would I gain anything by using rhyme?
Or aren't the bones enough I live to lime?

The Middleness of the Road

The road at the top of the rise
Seems to come to an end
And take off into the skies.
So at the distant bend

It seems to go into a wood,
The place of standing still
As long the trees have stood.
But say what Fancy will,

The mineral drops that explode
10 To drive my ton of car
Are limited to the road.
They deal with near and far,

But have almost nothing to do
With the absolute flight and rest
The universal blue
And local green suggest.

from *In the Clearing*

A Cabin in the Clearing

For Alfred Edwards

MIST. I don't believe the sleepers in this house
Know where they are.

SMOKE. They've been here long enough
To push the woods back from around the house
And part them in the middle with a path.

MIST. And still I doubt if they know where they are.
And I begin to fear they never will.
All they maintain the path for is the comfort
Of visiting with the equally bewildered.
Nearer in plight their neighbors are than distance.

10 SMOKE. I am the guardian wraith of starlit smoke
That leans out this and that way from their chimney.
I will not have their happiness despaired of.

MIST. No one – not I – would give them up for lost
Simply because they don't know where they are.
I am the damper counterpart of smoke,
That gives off from a garden ground at night
But lifts no higher than a garden grows.
I cotton to their landscape. That's who I am.
I am no further from their fate than you are.

20 SMOKE. They must by now have learned the native
tongue.
Why don't they ask the Red Man where they are?

MIST. They often do, and none the wiser for it.
So do they also ask philosophers
Who come to look in on them from the pulpit.

They will ask anyone there is to ask–
In the fond faith accumulated fact
Will of itself take fire and light the world up.
Learning has been a part of their religion.

SMOKE. If the day ever comes when they know who
30 They are, they may know better where they are.
But who they are is too much to believe–
Either for them or the onlooking world.
They are too sudden to be credible.

MIST. Listen, they murmur talking in the dark
On what should be their daylong theme continued.
Putting the lamp out has not put their thought out.
Let us pretend the dewdrops from the eaves
Are you and I eavesdropping on their unrest–
A mist and smoke eavesdropping on a haze–
40 And see if we can tell the bass from the soprano.

Than smoke and mist who better could appraise
The kindred spirit of an inner haze?

For John F. Kennedy His Inauguration

Gift outright of *The Gift Outright*
(With some preliminary history in rhyme)

Summoning artists to participate
In the august occasions of the state
Seems something artists ought to celebrate.
Today is for my cause a day of days.
And his be poetry's old-fashioned praise
Who was the first to think of such a thing.
This verse that in acknowledgment I bring
Goes back to the beginning of the end
Of what had been for centuries the trend;
10 A turning point in modern history.
Colonial had been the thing to be
As long as the great issue was to see
What country'd be the one to dominate
By character, by tongue, by native trait,
The new world Christopher Columbus found.
The French, the Spanish, and the Dutch were downed
And counted out. Heroic deeds were done.
Elizabeth the First and England won.
Now came on a new order of the ages
20 That in the Latin of our founding sages
(Is it not written on the dollar bill
We carry in our purse and pocket still?)
God nodded His approval of as good.
So much those heroes knew and understood–
I mean the great four, Washington,
John Adams, Jefferson, and Madison–
So much they knew as consecrated seers
They must have seen ahead what now appears:
They would bring empires down about our ears

30 And by the example of our Declaration
Make everybody want to be a nation.
And this is no aristocratic joke
At the expense of negligible folk.
We see how seriously the races swarm
In their attempts at sovereignty and form.
They are our wards we think to some extent
For the time being and with their consent,
To teach them how Democracy is meant.
'New order of the ages' did we say?
40 If it looks none too orderly today,
'Tis a confusion it was ours to start
So in it have to take courageous part.
No one of honest feeling would approve
A ruler who pretended not to love
A turbulence he had the better of.
Everyone knows the glory of the twain
Who gave America the aeroplane
To ride the whirlwind and the hurricane.
Some poor fool has been saying in his heart
50 Glory is out of date in life and art.
Our venture in revolution and outlawry
Has justified itself in freedom's story
Right down to now in glory upon glory.
Come fresh from an election like the last,
The greatest vote a people ever cast,
So close yet sure to be abided by,
It is no miracle our mood is high.
Courage is in the air in bracing whiffs
Better than all the stalemate an's and ifs.
60 There was the book of profile tales declaring
For the emboldened politicians daring
To break with followers when in the wrong,
A healthy independence of the throng,

A democratic form of right divine
To rule first answerable to high design.
There is a call to life a little sterner,
And braver for the earner, learner, yearner.
Less criticism of the field and court
And more preoccupation with the sport.
70 It makes the prophet in us all presage
The glory of a next Augustan age
Of a power leading from its strength and pride,
Of young ambition eager to be tried,
Firm in our free beliefs without dismay,
In any game the nations want to play.
A golden age of poetry and power
Of which this noonday's the beginning hour.

The Gift Outright

The land was ours before we were the land's.
She was our land more than a hundred years
Before we were her people. She was ours
In Massachusetts, in Virginia,
But we were England's, still colonials,
Possessing what we still were unpossessed by,
Possessed by what we now no more possessed.
Something we were withholding made us weak
Until we found out that it was ourselves
10 *We were withholding from our land of living,*
And forthwith found salvation in surrender.
Such as we were we gave ourselves outright
(The deed of gift was many deeds of war)
To the land vaguely realizing westward,
But still unstoried, artless, unenhanced,
Such as she was, such as she would become.

The Draft Horse

With a lantern that wouldn't burn
In too frail a buggy we drove
Behind too heavy a horse
Through a pitch-dark limitless grove.

And a man came out of the trees
And took our horse by the head
And reaching back to his ribs
Deliberately stabbed him dead.

The ponderous beast went down
10 With a crack of a broken shaft.
And the night drew through the trees
In one long invidious draft.

The most unquestioning pair
That ever accepted fate
And the least disposed to ascribe
Any more than we had to to hate,

We assumed that the man himself
Or someone he had to obey
Wanted us to get down
20 And walk the rest of the way.

Notes

The Pasture

Robert Frost insisted that this poem, with its relaxed, homely tone, should preface all the collected and selected editions of his poetry. Originally published in *North of Boston* and written to his wife Elinor, it can also be read as an invitation to the reader to enter the poet's chosen world: the farm with its pasture and cattle. When Eleanor Farjeon edited her selection of Frost's poetry (1960), Frost himself suggested the title, *You Come Too* (see Approaches, p.105).

2–4 **I'll only stop ... be gone long** note the references to time here – a major theme in Frost's poetry.

from A *Boy's Will*

Ghost House

This poem achieves its impact by striking the same note of unresolved mystery as Walter de la Mare's *The Listeners* and Rudyard Kipling's *The Way Through the Woods*. How can the speaker be dwelling in a lonely house which *vanished many a summer ago* (2)? Why is the speaker's heart *strangely aching* (11), and what is he referring to when he says *in view of how many things* (29)? Above all, who or what is the speaker – is he himself a ghost?

16 **whippoorwill** a nightjar.

25 **mar** spoil; the moss covers over the names on the gravestones under the tree.

26 **They are tireless folk** another puzzle – how does the speaker know things about the couple but yet not know who they are?

27 **close-keeping** keeping close to each other, or keeping themselves to themselves.

Waiting

The importance of dreams and dreaming as a way of approaching reality is a theme not only in this poem but also in *Mowing* (p.5) and *After Apple-Picking* (p.25). What does the subtitle, *Afield at dusk*, add to your understanding of the poem? Notice the roundabout way in which this poem arrives at its conclusion – how and why do the last two lines make you re-read the whole of the rest of the poem?

2 **haycocks** haystacks.

4 **late** recently.

5 **antiphony** literally, singing or chanting by two alternating choirs; here the last light of sunset on the one hand and the first light of the full moon rising on the other.

9 **the opposing lights** i.e. the sunset and the rising moon referred to in the previous stanza.

16 **pirouettes** turns circles, like a dancer.

17 **purblind** unable to see clearly.

18–19 **rasp/In the abyss of odor** a grating sound made by an animal (a fieldmouse?). Because it is dusk, the field has become indistinct – the speaker is aware only of the great space (*the abyss*) and the strong smell of the cut hay.

20 **advent** coming.

23 **the worn book of old-golden song** the book is presumably worn because it has been well-used, and has been well-used because it contains much-loved and familiar songs from the past. What is this image intended to convey at this point in the poem?

27 The person for whom this poem is intended as a gift once she reads it.

Mowing

This sonnet centres on the paradoxes expressed in lines 9 and 13: *Anything more than the truth would have seemed too weak* and *The fact is the sweetest dream that labor knows*. These paradoxes are central to Frost's poetry, and go a long way towards providing a key to his writing. His poems are often notable for a seemingly dogged determination to present the facts and details of a situation, no matter how trivial. By this means, rather than through generalization, Frost achieves an unusual degree of realism and challenges the reader to deny the *truth* (9) of what is being described. (For a discussion of Frost's treatment of nature and the imagination see Approaches, p.116.)

Mowing is the poem which Frost himself thought the most successful in his first volume, *A Boy's Will*. How do you think it compares with the other poems from his first book, selected here?

8 **fay** fairy, spirit.
10 **swale** a low-lying piece of ground.
12 **orchises** orchids.

The Tuft of Flowers

In this poem the speaker moves from a belief that everyone lives essentially alone (lines 8–10) to a conviction that '*Men work together ... Whether they work together or apart*' (39–40). How does he come to change his view so completely?

Frost's expression of delight in this poem at the work of the mower and what he learns from him can be compared with Wordsworth's poems *The Leech Gatherer* and *The Solitary Reaper*. Both of these poems pay tribute to the essential dignity and wisdom of those who undertake solitary manual work and thus achieve a close understanding of, and sympathy with, nature.

6 **whetstone** a stone on which the mower would sharpen the blade of his scythe.
12 **a bewildered butterfly** in what sense is the butterfly bewildered, and what significance does he have for the speaker?
23 **A leaping tongue of bloom** the *leaping tongue of bloom* is like a tongue of flame – perhaps an echo of the biblical story of the tongues of flame representing the Holy Spirit which descended on the Apostles at Pentecost (*Acts*, 2, 3). For the speaker (and for the butterfly) the tuft of flowers offers an epiphany, a moment of revelation and inspiration – *a message from the dawn* (30) – which brings him into closer harmony with nature.
33 **a spirit kindred to my own** the speaker feels such an affinity with the mower (even though he cannot see him) that from now on it is as if they are working in the field together.

In Hardwood Groves

This poem illustrates again Frost's ability to present simple but essential truths on the basis of his observation of nature. The symbolism suggests the theme of death and resurrection that appears elsewhere in Frost's poetry. Lines 7 and 8 suggest a childlike intuition of things as they really are. What is the impact and significance of the triple repetition of *must* (7–9)?

Title **Hardwood** i.e. deciduous trees which lose their leaves in autumn.

11 **However it is** whatever the situation may be. The speaker observes the cycle of death and re-birth (leaves falling, being turned to compost; shoots of new plants forcing their way up through the dead leaves). It is essential, the poem implies, to accept things simply as they are.

from *North of Boston*

Mending Wall

This is one of Frost's most anthologized poems but ironically many readers have mistakenly assumed that Frost's own philosophy of human relationships can be summed up in the statement '*Good fences make good neighbors*' (45) because the poem terminates on this line. In fact, the poem proclaims Frost's belief that walls (and barriers of any kind) *give offense* (34). Yet is his attitude as unambiguous as he implies? After all, it is the speaker who has agreed to assist his neighbour in mending the wall, rather than let it collapse altogether; indeed, he initiates the process. Perhaps the poem's power lies in the tension between the neighbour's conviction that barriers are a good thing and the narrator's more tentative position – *Something there is that doesn't love a wall* (1). What is achieved by calling the poem *Mending Wall* rather than 'Mending a Wall'?

The distinctive form of the poem (unrhyming, decasyllabic – ten syllable – lines) and the colloquial style of the monologue enable Frost to create a style and tone of voice which he will use frequently in later narrative poems (e.g. *The Death of the Hired Man*, p.11 and *Home Burial*, p.17). Inversions such as the opening line (*Something there is* ...) and his use of amplification to make clearer what he wants to say (e.g. *The gaps I mean,/No one has seen them made* ... [9–10]) all become characteristic features of Frost's style.

5 **The work of hunters is another thing** i.e. that doesn't love a wall.

24 **He is all pine and I am apple orchard** i.e. the two types of woodland on either side of the wall.

28 **Spring is the mischief in me** Spring makes me feel mischievous. Note that the speaker only wonders *If I could put a notion in his head* (29). The things he would like to say to his neighbour remain unsaid, so the rest of the poem becomes, in effect, an argument that the speaker has with himself.

40 **an old-stone savage** a primitive stone-age hunter. (Compare with *To an Ancient*, p.58.) Is the speaker making an ironic comment on his neighbour's primitive views of social relationships – hence line 41?

41 **He moves in darkness** the speaker feels that his neighbour has never developed his own philosophy of life; instead *He will not go behind his father's saying* (43). By contrast the speaker prefers to see barriers between people coming down not going up:

> Before I built a wall I'd ask to know
> What I was walling in or walling out,
> And to whom I was like to give offense. (32–34)

The Death of the Hired Man

This poem forms a striking contrast with *Home Burial* (p.17). Far from the sense of failure and breakdown in communication evident in the latter poem, *The Death of the Hired Man* centres on a relationship and on a home which is built on trust and affection. Nevertheless, even here there are tensions and disagreements between the husband and the wife. The central section of the poem (lines11–120) consists of a dialogue between Mary and Warren, a young couple running a farm, in which they discuss whether or not it would be appropriate to allow an old and by now incompetent farm hand to have his job back. Frost himself had had experience both of working as a hired man (a casual farm labourer) and of running his own farm. (See Approaches, p.103.)

30 **I'm done** I can't take any more.

58 **He ran on Harold Wilson** the hired man keeps talking about one of his former workmates.

63 **lay this farm** cut all the hay in the fields belonging to the farm.

73 **piqued** irritated, got under his skin.

83 He knew how to locate underground springs by water divining (dowsing) with a forked branch of hazel held in the hands.

98 **the fool of books** Silas believes that the boy (Harold Wilson) will look a fool if he spends his time reading rather than learning practical skills.

107 **the harplike morning-glory strings** morning glory is a climbing plant (often wild) whose tendrils here stretch right from the ground to the eaves of the house.

118 **Home is the place ...** the two definitions of home offered here and in lines 119–120 emphasize the value placed on home by Mary and Warren. The importance of home for them makes Silas's homelessness all the more poignant – hence the significance of Mary's comment *he has come home to die* (111).

134 **better than appearances** Silas's brother may be more kindhearted than his behaviour so far has suggested.

141 **kinsfolk** relatives, members of the same family.

160–163 It is characteristic of Frost that he focuses our attention (and Mary's) on an apparently trivial detail – whether or not a small cloud covers the moon as it sails past – while a momentous event (the death of the hired man) is being discovered by her husband.

162 **making a dim row** the moon, the cloud and Mary; they are only dimly seen because the cloud drifts across the moon.

Home Burial

On one level, this is one of the most clearly autobiographical of Frost's poems. His own son, Elliott, died in infancy in 1900, putting a severe strain on his relationship with his wife, Elinor. Frost's biographer, Lawrance Thompson (*Robert Frost*, p.116) speaks of the *unhealed wound of his grief over Elliott's death* and comments that *Inseparable from that grief was his puzzled awareness that his relationship with Elinor had not developed as he had hoped. Home Burial* is not simply autobiographical, however. From its opening words: *He saw her from the bottom of the stairs* (1), it presents an impersonal account of a failing relationship which is at once private and universal. The title refers not only to the child buried within sight of the house where its parents live but also to the sense in which both the man and the woman have been buried alive in a house and in a marriage from which they cannot escape. Their relationship has become a kind of living death. As in *Mending Wall* (p.9) the barriers created by the walls of the house are psychological more than physical.

12 **you must tell me, dear** the term *dear* seems so inadequate here (and again in line 42) as if to imply a failure of true feeling. The man is aware that his wife is troubled (as her behaviour on the stairs – *Looking back over her shoulder at some fear* (3) – implies) but is unable to find the right words or tone to reach her.

49–50 **A man must partly give up being a man/With womenfolk** what do these lines suggest about the man's experience of marriage? Do they affect your attitude towards him?

59–62 The husband's failure to understand the woman's grief and anger is expressed in the limp assurance that he is *not so much/Unlike other folks* (59–60) and his comment *I do think, though, you overdo it a little* (62).

89–90 The husband speaks here. Do you think that he seriously believes he is cursed, or is he trying to use sarcasm to make light of his wife's criticism of his apparent lack of feeling over the death of their child?

102–105 The woman's complaint about people's inability to face death and to sustain grief is the poem's central point:

> 'Friends make pretense of following to the grave,
> But before one is in it, their minds are turned
> And making the best of their way back to life
> And living people...' (102–105)

By saying *before one is in it* (103), the wife seems almost to imply her own death. On the other hand, these lines can be taken to imply the poet's attitude – that one has to accept and endure life. (Compare with '*Out, Out–*', p.34 where a similar theme is explored.)

116 What does the poem's abrupt and inconclusive ending suggest about the man's final understanding of his wife's feelings? Why does Frost frequently end his poems on an unresolved note like this?

The Black Cottage

The central idea of this poem is summed up in the paradoxical remark:

> 'For, dear me, why abandon a belief
> Merely because it ceases to be true.' (105–106)

Significantly, Frost turns a question into a statement by refusing to use a question mark. *The Black Cottage* is an important example of a narrative poem in which nothing actually happens, but a chance encounter or observation sets off a chain of thought or memory which becomes the poetic focus of Frost's imagination. From a negative opening (the abandoned cottage passed by chance) the poem moves slowly towards the conclusion that beliefs don't become invalid just because they go out of fashion:

> 'Most of the change we think we see in life
> Is due to truths being in and out of favor.' (109–110)

(For a further discussion of this poem, see Approaches, p.114.)

3 **tar-banded** tar is often used as part of the process of grafting fruit trees.
22 **A buttoned haircloth lounge** an upholstered settee.
24 **daguerreotype** an early form of photograph.
25 **to war** the American Civil War.

31 **Gettysburg or Fredericksburg** battles in the American Civil War, in 1863 and 1862 respectively.

41 **the considerate neglect** the old woman had taught her sons to be independent and did not mind that they rarely came back to visit her.

52–53 **Garrison/And Whittier** William Lloyd Garrison (1805–1879) and John Greenleaf Whittier (1807–1892) were leading figures in the movement to abolish slavery. Whittier was one of the most celebrated American poets of the Nineteenth century, best remembered today for his hymn 'Dear Lord and Father of Mankind'.

61 **That all men are created free and equal** a principle enshrined in the opening statement of the American Declaration of Independence (1776):

> We hold these truths to be self-evident, that all men are created equal and that they are endowed by their Creator with certain unalienable Rights, that among these are Life, Liberty and the pursuit of Happiness.

64 **Jefferson's** Thomas Jefferson, third American President and principal author of the Declaration of Independence.

68 **the Welshman** Jefferson claimed Welsh ancestry.

74 **latter** more recent.

88 **Creed** the statement of Christian faith, beginning: *I believe ...*

94 **"descended into Hades"** in the Creed, it is stated that Jesus *descended into hell.* In classical Greek mythology, Hades was the underworld.

119–124 The minister gives a detailed description of the desert island he would like to be monarch of: it would have *Scattered oases* (119), *Sand dunes* (120), and a child born in the desert there would find its body covered with grains of sand like a sprinkling of sugar. It would not be a hospitable island because the unfashionable *truths we keep coming back and back to* (114) are not the truths that are popular.

120 **tamarisk** a shrub found on seashores.

125 **clapboards** weatherboarding on the outside of the cottage.

After Apple-Picking

This poem, with its free verse-form, and its relaxed present tense, first-person description, takes Frost into a new direction which will be increasingly important in his poetry. Out of an intensely experienced and documented moment – looking out on the scene where the apple-picking has been happening – the poet is able to confront truths about himself and to ask questions about what lies beyond human experience. It is worth comparing this poem with Keats's *Ode to a Nightingale* to see how both writers present the idea of sleep and death. (See Approaches, p.100.) Note also the set of contrasts on which the poem is built: summer/winter, work/rest, effort/reward, sleep/wakefulness; how do these contrasts contribute to the poem's effect?

7 **Essence of winter sleep** the scent of the apples reminds the speaker of hibernation. Apples used to be stored in darkened rooms to preserve them for eating throughout the winter. These apple-stores had a very distinctive smell, rather like home-made dry cider.

10 **glass** ice.

22 **ladder-round** the rung on which he has been standing.

40 **woodchuck** a North American marmot, a squirrel-like rodent, also known as a groundhog.

41 **Long sleep** hibernation (an echo of line 7).

The Wood-Pile

Like *After Apple-Picking* (p.25), this poem is a meditation sparked by a seemingly trivial but vivid moment of experience. The pile of wood prompts the speaker to ask who could take such trouble over a task and then apparently forget about it:

> ... I thought that only
> Someone who lived in turning to fresh tasks
> Could so forget his handiwork on which
> He spent himself, the labor of his ax ... (34–37)

This poem has a bleakness emphasized at the start by Frost defining the place and the time of his experience as *the frozen swamp one gray day* (1) and culminating in the final image of the *slow smokeless burning of decay* (40).

23 **a cord of maple** a cord is a measure of cut wood, usually 128 cubic feet. The maple tree is found everywhere in North America and Canada (the maple leaf is the Canadian flag's emblem).

30 **Clematis** a climbing plant. During spring and summer clematis has a profusion of flowers, but in winter only its rope-like tendrils can be seen.

from *Mountain Interval*

The Road Not Taken

This poem, which Robert Frost claimed was about his friend, the poet Edward Thomas (see Approaches, p.107), is one of his best-known but most problematic poems. The title suggests that the subject is the road, the choice that the speaker did not take; the reader is apparently invited to ask whether the speaker regrets not having taken that road, or at least whether he regrets not having had the chance to explore both the roads which *diverged in a yellow wood* (1). However, the poem actually focuses on the road which *was* chosen, and the last line leaves open the question, exactly what is the *difference* (20) which choosing the less-travelled road has made. Thus, although the poem appears to be a brave assertion of the value of making unconventional decisions in life, it may also be read as a world-weary acknowledgement that you have to go on living with the consequences of your choices:

> I shall be telling this with a sigh
> Somewhere ages and ages hence: (16–17)

William H. Pritchard (*Frost: A Literary Life Reconsidered*, p.128) describes *The Road Not Taken* as *a notable instance in Frost's work of a poem which sounds noble but is really mischievous*. Ian Hamilton says:

> The air of lostness, of irretrievable error that hangs over the poem is a beguiling means of disguising its essentially inert bleakness. To Frost, it doesn't seem to matter much which road he took, or didn't take. It is that indifference which should have been the real subject of the poem. (*Robert Frost: Selected Poems*, pp.18–19)

The poem's unusually taut verse-form (nine-syllable lines, rhyming ABAAB) contrasts with its apparently casual tone (*perhaps the better claim* [7], *really about the same* [10], *I doubted if I should ever come back* [15]).

1 **yellow wood** i.e. in autumn. The paths are covered in fallen leaves (see line 12). Does this imply a choice made late on in life? (This would be appropriate if the poem is about Edward Thomas: Thomas had to choose in his late thirties whether to join up and fight in the First World War or whether to follow Frost's suggestion and emigrate to America.)

8 **wanted wear** lacked use; no one had walked this way lately so the grass had not been trodden down.

16 By using the future tense in this line, the speaker implies that his journey has not yet come to an end.

The Oven Bird

This poem, in the form of a rather unconventional sonnet, illustrates Frost's particular strain of pessimism. At the height of midsummer, the speaker is already anticipating autumn, *that other fall we name the fall* (9). The bird who sings is (according to Jeffrey Meyers, *Robert Frost, A Biography*, p.138) *an American warbler with an oven-shaped nest* and the question with which the poem ends (*what to make of a diminished thing* [14]?) invites the reader to accept that life itself, like the summer, is past its best. Alternatively, the question could be seen as a challenge: just as the bird – although he senses the onset of autumn – is able to fill the wood with his singing, how can we celebrate a life which is less than perfect? Frost explores the bird's ability to celebrate life as a temporary phenomenon and thereby to transcend it. (See Approaches, p.120.)

12 **he knows in singing not to sing** although the bird sounds cheerful he knows that the future is nothing to sing about.

14 **a diminished thing** something which is no longer as fine or impressive or desirable as it used to be.

Birches

Drawing on his own childhood memories, Frost creates an image (birch trees bent almost to the ground by a child who climbs and swings on them) that sums up all the ambiguity which as an adult he feels about life:

> I'd like to get away from earth awhile
> And then come back to it and begin over. (48–49)

Look carefully at the last six lines of the poem: what meaning and feeling can you discern behind the line *One could do worse than be a swinger of birches* (59)? Robert Frost wrote that *Birches is two fragments soldered together so long I have forgotten where the join is.* Can you locate it?

5 **Often you must have seen them** notice how the speaker draws the reader into the poem. This single reference to *you* turns the poem from a private meditation into a monologue directed at a listener. This idea is reinforced in line 21: *But I was going to say when Truth broke in.*

44 **life is too much like a pathless wood** this image (is it a cliché?) is one which was shared both by Frost and by Edward Thomas. (See Approaches, p.109 and Thomas's poem *Old Man* in the Appendix, p.144.)

56 ***Toward*** why does Frost put this word in italics?

The Cow in Apple Time

In this poem the poet seems to be amused by the drunken cow but characteristically the last line strikes a very different note: the cow is no longer an image of fun but of scorn and decay. How does the unusual rhyme scheme of this poem enhance and emphasize the poet's point of view?

3 **wall-builders** see *Mending Wall* (p.9).

4 **pomace** the pulp of apples after they have been crushed for cider making.

8 **windfalls** apples that have been blown off the tree or have fallen of their own accord. (See *After Apple-Picking*, p.25.)

10 **knoll** a small hill.

11 Look carefully at this last line; why does the poem end with these images of uselessness?

An Encounter

Out of an almost comical insight – that a telegraph pole is both *a resurrected tree* (12) and *A barkless specter* (14) – and an equally comical encounter (the speaker talks to the telegraph pole about the destination of the messages being carried along its wires), Frost creates a poem in which the speaker acknowledges his own apparent lack of direction (though not entirely his lack of purpose):

> 'Me? I'm not off for anywhere at all.
> Sometimes I wander out of beaten ways
> Half looking for the orchid Calypso.' (22–24)

What is the effect of the intricate irregular rhyme-scheme of this poem?

1 **'weather breeder'** perhaps the sort of atmosphere which seems to be building up to a thunder storm and a change in the weather.

6–7 **and weary and overheated,/And sorry** the repetition of *and* emphasizes the speaker's sense of discomfort and annoyance with himself.

7 **sorry I ever left the road I knew** see *The Road Not Taken* (p.28).

12 **a resurrected tree** this image, amplified by the next line (*A tree that had been down and raised again*) deliberately borrows the language of Christianity in echoing the idea of Christ's death and resurrection.

24 **Calypso** in Greek mythology the queen of an island (probably Gozo, near Malta) where Ulysses was shipwrecked. She kept him there for seven years, promising him perpetual youth if he remained faithful to her. Calypso is also the name of a rare orchid found in North America, also called Lady's Slipper or Moccasin.

'Out, Out–'

The impact of this poem springs from two shocks: first, the loss of the hand in the accident with the saw, and then the death of the boy. These events seem the more shocking because of the impersonal tone of the narrative (*So. But the hand was gone already* [27]) and by the apparent indifference to tragedy reflected in the final sentence: *And they, since they/ Were not the one dead, turned to their affairs* (33–34). Note the use of half-rhyme in this poem – with its lack of resolution. What does it contribute to the impact of the writing?

Title **'Out, Out–'** see *Macbeth: Out, out brief candle!*. A line from the soliloquy, spoken by Macbeth, after he has been told of the death of his wife. He speaks of the apparent pointlessness and brevity of life.

12 **when saved from work** ironically the half-hour of free time would have saved his life.

23–24 **big boy/Doing a man's work, though a child at heart** how well does this reflect Frost's ambiguous feelings about himself and his life?

28 **the dark of ether** the anaesthetic, but also anticipating the darkness of death.

33–34 Compare the apparent indifference of the adults here with the complaints of the wife in *Home Burial* (p.17) of the callous attitude of people towards the death of her child. What does the reaction of the adults to the child's death in this poem tell us about these people and their lifestyle?

The Sound of Trees

By contrast with Frost's more usual decasyllabic (ten-syllable) lines, the short lines of *The Sound of Trees* give the poem an unsettling tone which is reflected in the speaker's certainty that one day he will *make the reckless choice* (20) even though he does not yet know where that choice will lead him. Some of the deliberately awkward phrases (e.g. *They are that that talks of going* [10], and *tossing so as to scare/The white clouds over them on* [22–23]) help to reinforce the sense of dislocation. Why does the speaker include a second person in lines 2–5 (*Why do we wish to bear ... our dwelling place?*) when, by line 16, he is speaking as if he is an isolated figure (*my head sways ... I shall set forth for somewhere*)? The poem seems to end with the speaker's final, almost unnoticed, disappearance:

> I shall have less to say,
> But I shall be gone. (24–25)

3 **the noise of these** is it just the trees that will drive the narrator away with their incessant noise or are they a metaphor for something else? Words such as *bear* (2), *suffer* (6) and *scare* (22) all suggest a pain that has a human source.

from *New Hampshire*

The Ax-Helve

Although the title suggests that this poem has as its central theme or image the handle (helve) of an axe, Frost presents several topics here: in particular, he is concerned with the way people (in this instance, neighbours) create relationships while at the same time maintaining a sense of distance. (Compare with *Mending Wall*, p.9.) The poem also focuses on the question of education and what rights being educated confers on people. It is as if the speaker himself is surprised to find such a topic emerging in the conversation he has when he visits his neighbour: *Do you know, what we talked about was knowledge?* (82).

The Ax-Helve is a rare example of Frost's use of dialect in a poem – in this case the French-English idiom of the neighbour Baptiste: the awkwardness of his speech (*She ain't spick too much Henglish – dat's too bad* [54]) reflects his own awkward but persistent personality. (See Approaches, p.103.)

2 **alder** a tree of the same variety as the birch, with catkins and serrated leaves.

5 **as I say** note the relaxed tone achieved here by the use of near repetition so early in the poem.

24 **the handle's long-drawn serpentine** the distinctive curving shape of the axe handle.

30 **good hick'ry** hickory is a tough, heavy wood from which walking sticks and tool-handles are often made.

38–43 This very complex sentence (itself almost serpentine like the handle of the axe) reflects the difficulty the two men have in weighing each other up. The speaker isn't going to let himself be worried by the position in which he has been placed. He will have to judge whether Baptiste's expert knowledge about axes gives him a special claim to be respected as a neighbour.

44 **Yankees** a slang term for the inhabitants of New England or one of the northern United States. Frost himself came from this area.

69 **whipstock** the handle of a whip.

76 **chafed** rubbed.

78 **the eyehole in the ax-head** the hole into which the handle of the axe is fitted.

83 **on his defense** on the defensive.

90–93 Another 'serpentine' sentence. As in lines 38–43 above, here the speaker has to make up his mind about what Baptiste wants from him as a neighbour: 'Did Baptiste want me as a friend partly because he could trust me to decide whether or not he was well enough educated to be entitled to have his own views on how his children should be educated?'

97 **The snake ... in the Garden** a reference to the serpent which tempts Eve to eat the apple from the tree of the knowledge of good and evil, as described in the Bible (*Genesis*, 3, 1–16). The serpent was often depicted (e.g. in stained glass windows) as standing erect or twining around the trunk of the tree – hence it *stood up for evil* (97).

To E.T.

Robert Frost's friendship with Edward Thomas was one of the most significant of his life. (See Approaches, pp.107–110.) Thomas was killed in action during the Battle of Arras in April 1917.

This poem, addressed to Thomas himself, is strikingly formal (by Frost's standards) in its structure and diction: calling Thomas *a soldier-poet of your race* (8) is a most un-Frost like phrase. As in *The Road Not Taken* (p.28) Frost meditates here on choices and opportunities missed. Here too Frost is not afraid to couch difficult ideas in complex and obscure statements. Lines 9–12, the central stanza of the poem, seem to defy interpretation. What Frost means by *The Victory for what it lost and gained* (12) is obscured by the syntax of the sentence in which it appears. By contrast, the poem ends with a touchingly simple final question which is intended to reflect the delight that Frost took in Thomas's friendship and in their shared experience of poetry.

Compare with *Easter Monday* by Eleanor Farjeon (Appendix, p.149). Why do you think Frost identifies Thomas only by his initials in the title of this poem?

3 Interestingly, Frost presents himself here as an effigy, a carved stone figure on top of a tomb, covered by the wings of doves, the symbol of peace.

13 Thomas was killed by a shell during the Battle of Arras (*Vimy Ridge* [14]).

18 **The foe thrust back unsafe beyond the Rhine** i.e. into Germany. There was continuing anxiety in Europe and America during the 1920s and 1930s that Germany (*The foe* – note again the formal, poetic language) would provoke another war.

Stopping by Woods on a Snowy Evening

Along with *The Road Not Taken* (p.28) this is one of the best known of Frost's poems. It is also one of the most troubling and some critics have suggested that its underlying theme is suicide. The title seems to suggest the caption for a picture or photograph, as if the poem is a snapshot in words. And yet Frost is not simply concerned to paint a scene, for behind the images (woods, snow, midwinter – *The darkest evening of the year* [8]) and the almost child-like rhythms and rhyme scheme lies the idea of a choice which is dangerously tempting. In the final stanza *The woods are lovely, dark, and deep* (13) the speaker is held back from entering them by the promises, the commitments, which will force him to continue his journey. He may literally have *miles to go* (15) before he can rest, but the repetition of the line may also suggest that he realizes he has to live out the rest of his life before he can die (*sleep* [15–16]).

2 **His** as with *Whose* in line 1, the speaker does not identify the owner of the woods. At the outset, therefore, a note of mystery and anxiety is created.

5 **My little horse** what use is made of the horse and of the owner of the woods in the poem?

8 **The darkest evening of the year** the winter solstice (21 December). Compare with John Donne's poem *A Nocturnal upon St Lucie's Day*, which is also set on the same day.

13 Some manuscript versions of the text of this poem omit the comma after the word *dark*. What difference would this make to a reading of the line and, perhaps, of the poem as a whole?

Two Look at Two

This poem is set in Frost's familiar landscape: woods, paths, walls all feature not as background but as setting for an encounter between a human couple and a doe and a buck. The meeting forces the man and woman to reassess and perhaps to reaffirm the importance of their relationship. The querulous tone in which, the poet suggests, the buck addressed the human couple: '*Why don't you make some motion?/Or give some sign of life? Because you can't.*'(32–33) is reminiscent of the anguished wife in *Home Burial* (p.17). The ending of the poem (lines 39–42) suggests a kind of epiphany (a moment of insight or intuition) in which the true nature of their relationship (and of man's relationship with nature) is revealed to the man and woman.

6 **washout** a geological technical term. Frost is referring to a narrow river channel that cuts into pre-existing sediment. Here, the path is partially washed away.

27 **A buck from round the spruce** compare with *The Most of It* (p.52), lines 10–20 where another buck makes a dramatic appearance.

Gathering Leaves

Frost's use of a very compact verse-form means that the lines of this poem are presented as a series of aphorisms (statements of general truths). The physical description of gathering leaves (stanzas 1–3) leads on to a universal question about how to put a value on work that has been done:

I may load and unload
Again and again
Till I fill the whole shed,
And what have I then? (13–16)

10 **Elude** avoid, escape.

from *West-Running Brook*

Tree at My Window

In poems such as *Birches* (p.30) and *The Sound of Trees* (p.35) Frost has produced meditations prompted by trees; in this poem the speaker addresses a tree directly. The writing is notable for its strikingly original description of the tree (*Vague dream-head lifted out of the ground,/And thing next most diffuse to cloud,* [5–6]) and for the sense of shared identity that the speaker feels. The poem is also distinguished for its careful modulation (deliberate variation) of tone: look particularly at the effect achieved by the use of the short fourth line of each stanza.

2 **sash** the lower portion of the window.

7 **your light tongues talking aloud** the leaves rustling in the wind.

13 Compare the way the speaker here manages to identify with the tree (while carefully insisting on the difference between them) with the way the speaker compares and contrasts himself with the leaves in *A Leaf-Treader* (p.49). Note also the wit of the analogy.

Acquainted with the Night

The use of repetition in this poem, both of whole lines and of phrases, helps to establish an almost chant-like effect which is emphasized by the inter-linking rhyme scheme. (See *Preludes* and *The Love Song of J. Alfred Prufrock* by T.S. Eliot and *As I Walked Out One Evening* by W.H. Auden for other contemporary poems where the speaker has been 'acquainted with the night' and the setting is a bleak urban night landscape.) How does Frost establish the tone of this poem, and what clues does the tone offer about the reasons for the speaker's solitary walking?

12 **luminary** not just lit up but light-giving.

A Soldier

Written in the form of a sonnet, this poem (rather like *To E.T.*, p.40) adopts a formal, impersonal note. How does his use of diction enable Frost to achieve this note? From the opening line with its description of the soldier as *that fallen lance* (an example of metonomy, where the whole being - the soldier - is represented by one of his parts - his fallen lance) onwards, Frost develops the idea of the soldier and his sacrifice as somehow greater than ordinary human endeavours: *Our missiles always make too short an arc* (8).

7 **as fitted to the sphere** the sphere here refers to earth; as men and women we take too limited (earth-bound) a view of life and its potential.

12–14 Frost ends the poem with the assertion (*But this we know* [12]) that at the moment of death, the soldier's soul escapes the limits of the body and of this world. This is one of the rare occasions in Frost's poetry where the idea of an afterlife is clearly implied.

from *A Further Range*

Desert Places

This poem should be compared with *Stopping by Woods on a Snowy Evening* (p.41) and with *Tree at My Window* (p.45). The sense of physical and emotional isolation (*I am too absent-spirited to count* [7]) is strongly emphasized by the image of the falling snow (A *blanker whiteness* [11]). However, the speaker is not frightened by isolated places but by his sense of spiritual emptiness. The final stanza's insistence on the *empty spaces* (13) magnifies the speaker's fear of his own inner loneliness:

> I have it in me so much nearer home
> To scare myself with my own desert places. (15–16)

Thus the natural scene becomes a symbol of a state of mind – a common use of symbolism in Frost's poetry.

5 **The woods around it have it – it is theirs** the first *it* refers to the *field* (2); the repeated *it* – *it* emphasizes how the snow is becoming part of the landscape.
8 **includes** envelopes or embraces.
11 **benighted** literally, overtaken by darkness; metaphorically, morally and spiritually empty.
16 **scare** note the choice of a colloquial term here. What does it contribute to the poem that an alternative word such as 'terrify' might not have done?

A Leaf-Treader

Leaves are an important recurring image in Frost's poetry. (See Approaches, p.117.) Compare this poem with *Gathering Leaves* (p.44); in what ways are the final stanzas of each poem similar to, and different from each other?

2 **mired** covered in mud.
4 **the leaves of another year** having trodden down the leaves into the compost heap (*mired* [2]), the speaker – though exhausted – feels he can say that he has safely survived to the end of another year. The note here is of relief rather than achievement.
9 **the fugitive in my heart** that part of my personality that longs for escape.
leaf to leaf compare this image with the image of the tree in *Tree at My Window* (p.45).
12 **Now up, my knee** the poem ends with an image designed to emphasize the effort that will be needed to survive (*to keep on top of*) the coming year. Note that the final image is not of leaves but of snow threatening to overwhelm the speaker.

Neither Out Far nor In Deep

Like *The Sound of Trees* (p.35), *Stopping by Woods on a Snowy Evening* (p.41), *Gathering Leaves* (p.44), and *The Middleness of the Road* (p.59), this poem partly relies for its effect on the apparent simplicity and repetitiveness of its form and rhyme scheme. What effects are created by the use of short lines (usually endstopped - except for lines 7 and 15) and by the succession of very direct statements, culminating in a single question? (See Approaches, p.127.)

They cannot look out far.
They cannot look in deep.
But when was that ever a bar
To any watch they keep? (13–16)

There Are Roughly Zones

This poem appears to contain a statement of Frost's pragmatic philosophy of life. It is embedded in a question the speaker asks about why man never accepts that there should be limits to human aspirations and expectations:

Why is his nature forever so hard to teach
That though there is no fixed line between wrong and right,
There are roughly zones whose laws must be obeyed? (11–13)

The irregular long lines and rhyme scheme of the poem suggest a relaxed and almost patient teasing out of the problem, even while the storm is threatening to blow down the house and destroy the peach tree. (See Approaches, p.125.)

3 **the house has long been tried** storms have threatened to blow down the house many times in the past. It has always managed so far to survive.

6 **It is very far north, we admit** what point is Frost bringing out in this line? (Look at the last four lines of the poem.)

15 **more than a little betrayed** the speaker and his companion feel it unfair that the northwest wind should blow so strongly at the same time as the temperature drops so sharply. The tree has little chance of survival.

21 **this limitless trait in the hearts of men** this final line implies that Frost (up to a point at least) approves of the fact that the human spirit always wants to go further. There are only *roughly* (13) bounds or *zones* (13) within which man *can stay confined* (8).

from *A Witness Tree*

The Most of It

By contrast with *There Are Roughly Zones* (p.51) and *Two Look at Two* (p.42) this poem is riddled with anti-climaxes which seem to imply a strong sense of frustration and disappointment. The poem focuses on the figure of an isolated man living beside a lake. He cries out in frustration against the lack of *counter-love, original response* (8) in his life, and, in doing so, he highlights his own loneliness.

As with *The Oven Bird* (p.29), this poem challenges the reader to accept that life seems to give less than it promises. Is this inconsistent with the views apparently expressed in other of Frost's poems, or does this attitude lie at the heart of his philosophy?

The final statement (*and that was all* [20]) suggests that the title of the poem has to be taken ironically. How do the insistently regular rhyme scheme and iambic metre determine your interpretation of the poem's ending?

1 Significantly, this poem has a third-person perspective (*He*), which enables the narrator to hint that perhaps he himself does not share the views or the frustration of the man who *would cry out on life* (6).

7 **copy speech** exact repetition of what has just been said; an echo.

8 **counter-love** an equal but different response.

10 **embodiment** his words given bodily form so that they crashed into the rocks like a living creature.

11 **talus** the sloping mass of fragments of rock at the foot of the cliff.

16 **As a great buck it powerfully appeared** compare with *Two Look at Two*, lines 26–42 (p.42).

A Considerable Speck

Title **Considerable** worthy of consideration; also, of a certain size. Although the speck is *Microscopic*, the speaker pays it the compliment of taking it seriously, taking pity on its fear and admiring its intelligence.

5 **period** a full stop; here, literally a blob of ink.

24–25 The speaker says that he does not believe one should be forced into claiming to love everything and everybody.

32–33 The ironic note with which the poem ends (with the implication that most of what is written or printed is mindless and therefore worthless) is accentuated by the poet's use of rhyming couplets here as throughout the poem.

from *Steeple Bush*

A Young Birch

Although the tone of this poem is, on first reading, more casual than the tone of a poem such as *Birches* (p.30), Frost still sets out to celebrate the beauty of the tree in its own right. As the last sentence of the poem implies, the young birch exists simply to give pleasure to those who look at it. Frost insists that no other justification for its existence is needed.

5 **To double day** the gleaming whiteness of the bark will double the lightness of the day.

9 **Relying on its beauty** the speaker suggests that the tree is self-consciously aware of the effect its beautiful appearance has on those who see it.

11 **someone reminiscent** the poem moves forward as the speaker imagines a future time in which somebody will remember having spared the young birch when he was *cutting brush along the wall* (12). The young birch has been spared in the same way as the *leaping tongue of bloom* (23) in *The Tuft of Flowers* (p.6).

16 **bole** the stem or trunk of a tree.

19 **And zeal would not be thanked** someone who was over-conscientious and cut down the tree would get no thanks from anyone else for having done so.

20 **When you were reading books** what is the effect of introducing *you* into the poem at this point? Is the speaker simply referring to himself? In any case, the line seems to suggest that the tree, by simply standing as it does, is engaged in an activity just as worthwhile as reading or travelling.

An Unstamped Letter in Our Rural Letter Box

This poem teases in several ways. It is presented in the form of a letter addressed to the owner of a property by a tramp who has spent the previous night camping in the owner's pasture. Yet this tramp uses strangely formal language (*I elected to demur* [12]) and is happy to quote Latin in order to produce a final rhyming couplet. The letter apologizes not for the fact that the tramp has trespassed, but that he might *seem to boast* (39). What he boasts of is an experience he had during the night: two stars *having coalesced* (26) shot through the sky like *The largest fire-drop ever formed* (25), and this prompted a coming together of two memories in the mind of the speaker. The insight that followed was so powerful that: *for a moment all was plain/That men have thought about in vain* (36–37). What the insight was, the speaker does not say. Indeed, he implies that the owner of the property may well have had similar insights himself. He concludes by suggesting that it was:

> ... partly to compel
> Myself, *in forma pauperis*,
> To say as much I write you this. (50–52)

What other reasons he may have had for writing the letter, again he does not say. If the poem has a moral (and its tone becomes increasingly didactic) it is summed up in the line *Each knows his own discernment best* (43).

Compare the way Frost builds up pairs of contrasts (tramp/scholar, labourer/farm owner, educated/uneducated, etc.) in this poem with other poems where he uses the same technique (e.g. *After Apple-Picking*, p.25). Do you think these contrasts might reflect some of the contradictions Frost is trying to explore in his own personality?

8 **the pip of spades** the ace of spades. The playing card image is continued in the next line with *a suite of glades* (9).

12 Frost seems to be using the word *demur* here not in its usual sense of 'to make an objection' but to mean 'to stay' (as if from the French *demeurer* – 'to live').

13 **juniper** an evergreen shrub or tree.

26 **coalesced** having apparently come together to form one star, they went *streaking* (27) (like *molten* [27] metal) through the sky, like a shooting star.

34–35 **quivered ... lipped/Together ... together slipped** why do you think Frost uses sexual images here to describe how the two memories 'coalesce' like the two stars?

38 **my involuntary host** the person to whom the letter is addressed was an involuntary host because he had had no say in whether or not the tramp could spend the night on his property.

42 **Heaven showed your guest** does this reference to *Heaven* (and the other reference in line 30) suggest that the tramp sees some guiding hand behind the moments of outer and inner illumination he experienced?

51 ***in forma pauperis*** in the likeness of a pauper or tramp. Does this imply that the speaker is admitting he is not really a tramp at all?

To an Ancient

Reflecting on the discovery of traces of a prehistoric man, Frost is forced to ask the question, will my poetry make it any more likely that I myself shall be remembered in time to come?

2 **The one you made** this refers to the *eolith* (13), presumably a once-standing stone which has been found *in the delta of a brook* (5). The delta is where the brook widens as it runs into a larger river.

10 Archaeologists excavating the human remains of this prehistoric man can estimate his date and discuss his *probable brute nature* (11); however, beyond this nothing about the man can be known – how much more, the poem implies, does one need to know?

13 **eolith** a stone from the eolithic period, which preceded the paleolithic or old stone age. *Eos* (Greek) means 'dawn'.

17 **by using rhyme** i.e. by writing poetry.

18 **I live to lime** eventually my bones will decay back into earth. *Limus* (Latin) means 'mud'.

The Middleness of the Road

Once again, the theme of this poem is human limitation. The road may seem to go on indefinitely until it disappears into the sky (*The universal blue* [15]) or the woods (*local green* [16]) but the speaker's car is useless except on the road. Frost makes it clear that the sky and the woods for him represent the finality of death (*the absolute flight and rest* [14]). In what ways, therefore, can the poem's title be interpreted?

8 **Fancy** the imagination. Why is the capital letter used?

9 Here the poet is referring to the combustion (*explode*) of the fuel (*mineral drops*) which drives the engine of his car.

from *In the Clearing*

A Cabin in the Clearing

Although many of Frost's poems feature conversation, this is rare in being set out as a scripted dialogue between two speakers. The theme of the poem, that human beings spend much of their lives lost and uncertain but still capable of self discovery, is central to Frost's work. The key lines are spoken by Smoke:

> If the day ever comes when they know who
> They are, they may know better where they are. (29–30)

10 **wraith** ghost.

21–22 If the *Red Man* (21) represents the wisdom of the traditional inhabitants of the place, who is being criticized here – the settlers who (though they *must by now have learned the native tongue* [20]) are not able to understand what is said to them, or the Native American whose message is too obscure?

33 **too sudden to be credible** *sudden* here suggests 'recent' or 'new', as if the settlers (the builders of the log cabin) are such newcomers that they cannot yet be taken seriously by those – the guardians of the place – who have always been there.

38 **unrest** although the occupants of the cabin are *sleepers* (1), they are unable to rest peacefully because they cannot resolve

their *daylong theme* (35). Frost implies that fundamental questions about identity and the meaning of life trouble us all.

40 **the bass from the soprano** literally, the low note from the high note; here, perhaps, referring both to the voice of a man and a woman and reflecting the deeper and the higher (superficial) problems with which the settlers are having to wrestle. (Compare the *inner haze* [42] and the outer haze.)

For John F. Kennedy His Inauguration

(Gift outright of *The Gift Outright*)

Frost was invited to take part in President Kennedy's inauguration in 1960, an honour which made him in effect the unofficial Poet Laureate of America. He recited from memory his poem *The Gift Outright*, having prefaced it with the couplets that speak optimistically (naively?) about *A golden age of poetry and power* (76).

2 **the august occasions of the state** formal state ceremonies such as the inauguration of a new president.

4 **my cause** i.e. poetry.

11 **Colonial** i.e. in favour of a policy of colonizing the New World discovered by Christopher Columbus at the end of the Fifteenth century.

16 France, Spain and the Netherlands had all colonized parts of the New World; gradually they were displaced by *Elizabeth the First and England* (18).

20 **the Latin of our founding sages** *E Pluribus Unum* – 'Out of Many, One'; the inscription on American dollar bills. *Our founding sages* are the wise men (presidents of the United States) named in lines 25–26.

25–26 **Washington,/John Adams, Jefferson, and Madison** the *great four* (25) presidents who laid the foundations of the modern United States of America.

27 **consecrated seers** men with the gift of foresight, who had acquired almost the status of holy leaders.

30 **our Declaration** the American Declaration of Independence (1776). (See *The Black Cottage*, p.21, line 61.)

31 Is this America's goal, or its destiny?

36 **wards** in legal terms, a ward is a child for whom someone else (another adult or the courts for example) takes responsibility when their own parents are unable to look after them properly. Here, the reference is to emerging nations of the world who look to the USA for protection and support.

46 **the twain** the Wright brothers, pioneering American aviators.

51 Here Frost is referring to the whole past history of America, which he describes (perhaps tongue-in-cheek) as *revolution and outlawry* – a mixture of high and low political and social lawlessness.

55 Frost was a strong supporter of John F. Kennedy's campaign for the presidency. He saw Kennedy's election as heralding a new era which would see a return to the idealism of the early American puritan settlers, after the materialism of the Eisenhower presidency (1952–60).

71 **a next Augustan age** referring back to the golden age of the Roman Empire under Augustus.

The Gift Outright

1 **The land was ours** Frost claims that America belonged to the Americans (i.e. the American settlers) before they had obtained independence from Great Britain.

4 **In Massachusetts, in Virginia** early British colonies on the east coast of America.

12 **we gave ourselves outright** unconditional submission to the land. This implies not just a complete declaration of loyalty to America as a country, but literally and physically to *the land vaguely realizing westward* (14), the territory that was opened up by the frontiersmen.

The Draft Horse

One of Frost's last poems, *The Draft Horse* is a characteristically unsettling piece of writing. The account of the attack on the horse, delivered in completely unemotional and unembellished tones, is macabre enough; the apparently unquestioning way in which the speaker and his companion accept what has happened leaves the reader feeling disorientated. Is this poem intended to imply that Frost himself believes in a capricious Fate? If so, how would you interpret the buggy and the horse here? Why is the buggy *too frail* (2) and the horse *too heavy* (3)? The pair in the buggy are described as *the least disposed to ascribe/Any more than we had to to hate* (15–16). What does this mean and how does it affect the ending of the poem? In the final stanza, the speaker suggests that the riders will *get down/And walk the rest of the way* (19–20); why is this a surprising and powerful ending to *The Draft Horse?*

Title **Draft Horse** a large horse used for pulling heavy loads.
12 **one long invidious draft** one straight, indiscriminate swathe, like a ploughed furrow.

Approaches

Introduction

Putting the Poetry First

The best way to approach Robert Frost's poetry is through the poems themselves. This may seem too obvious to need saying, but it is easy to get sidetracked by discussions about Frost's career, his reputation, the opinion of the critics, etc. Once this happens, you will stop reading the poems for their own sakes and start instead to look for clues about Frost's personality or for reasons to support a particular critical view of his work. These things are important (and it will certainly be important for you to weigh up different attitudes to Frost's poetry and then to reach your own conclusions) but what matters most is your response to the poems themselves, based on a careful and open-minded approach to what Frost actually writes.

You will, however, need to know something about the facts of Frost's life, and the Chronology on page 130 will give you basic information. Some biographical background has already been offered in the Notes (see, for example, the Notes on *Home Burial*, p.73 or *For John F. Kennedy His Inauguration*, p.95); you will also find more material later on in these Approaches (see the section, Frost's Life, p.103). The information that is given, however, is simply there to help you to appreciate the poetry more fully.

You should also be mindful of the context within which Frost's poetry was written and may be understood. One of the questions that his writing raises is, 'Is there anything about this poetry which makes it distinctively American?' It may be a matter of setting and theme: poems set in a particular American landscape may have a special American atmosphere, but of course it is quite possible for a non-American writer to write a poem with an American setting (and vice versa: T.S. Eliot was an American poet who wrote memorably about London in *The Waste Land*). Frost is very much concerned with the environment and with the idea of living close to the land; he rarely sets his poems in cities or writes about urban culture or values. In this respect, he may seem to appeal to a particularly American 'frontier spirit', and poems such as

Birches (p.30), *The Ax-Helve* (p.36) or *A Cabin in the Clearing* (p.60) may appear to have a distinctively American atmosphere. On the other hand, there are few countries on earth with a greater diversity of people, cultures, landscapes, and traditions than the United States; is it possible that any one poet could be genuinely representative of such diversity?

You may be familiar with the idea of 'the American Dream', that confusion of aggressive material ambition on the one hand with individual freedom, the simple life, physical prowess, and the great outdoors on the other. Critics have argued over whether Frost's poetry offers a criticism of the American Dream or an endorsement of it.

Does Frost speak for traditional American values? Are the virtues that he seems to praise in his poetry – self-sufficiency, closeness to the physical and natural world, stubbornness, and stoicism (accepting fate with a shrug of the shoulders rather than railing against misfortune) – the virtues that many American readers would like to see in themselves, and which have led to Frost's popularity throughout the century among ordinary readers (if not among critics and other poets)? These are key issues, and your response to them must be based on a knowledge of the poems themselves.

After all, Frost the American poet may not be the same as Frost the universal poet. Often his poetry is concerned with choices, or with moments of insight and understanding which transcend any national boundaries. Two of his most famous poems, *The Road Not Taken* (p.28) and *Stopping by Woods on a Snowy Evening* (p.41) have themes and settings which cannot be confined to anything that is only – or even primarily – American. Here the symbolic settings (a crossroads, a dark wood) are so familiar as almost to be clichés, but Frost treats them in a powerfully direct way and gives them a significance that is both personal and universal:

> Two roads diverged in a wood, and I—
> I took the one less traveled by,
> And that has made all the difference. (18–20)
>
> (*The Road Not Taken*, p.28)

> The woods are lovely, dark, and deep,
> But I have promises to keep,

And miles to go before I sleep,
And miles to go before I sleep. (13–16)
(*Stopping by Woods on a Snowy Evening*, p.41)

These two quotations illustrate Frost's gift for investing very familiar phrases (*all the difference* [20], *miles to go* [15]) with a significance which we can feel but which is hard to define; in this way they show the universality of his writing. It would be possible to overlay an interpretation of each poem with a discussion of the circumstances in which it was written, with an account of Frost's state of mind at the time of composition, with speculation about the private meaning intended in each line. However, when a poem is published, it is released from the poet's own controlling presence, and every reader is free to respond to it in his or her own way.

A poem such as *After Apple-Picking* (p.25) describes very vividly what it feels like to have spent all day up a ladder picking apples; it then goes on to describe the drowsiness (halfway between waking and sleeping) that comes over the exhausted apple-picker, the speaker in the poem. It does not matter if one has never been up a ladder to pick an apple in one's life, the poem can still make a direct appeal: its descriptions are so precisely rooted in close observation and memory of the actual experience, and the questions the poem raises (about the nature of human desires and the gap between hopes and realities) are questions that everyone has to confront. When we read a poem we do not simply interrogate the text to tease out its meaning, we start to interrogate ourselves and to test our own understanding in the light of the poem we are reading.

Activity

Read *Ghost House* (p.2) and *A Cabin in the Clearing* (p.60). The first of these poems was written right at the start of Frost's career, while the second was written near the end. Almost fifty years separates their dates of publication. Compare the two poems to see whether they have any distinctive features in common. Does the fact that they are written by an American poet make any difference to the way you read them?

Discussion

Both poems are about ghosts or spirits, and about the importance of belonging. The first has a first-person narrator, while the second is a dialogue spoken by two *guardian* (10) spirits. In the first poem, the house *vanished many a summer ago* (2), but the cabin in the second has been there so long that the settlers have had *To push the woods back from around the house* (3). In their different ways, both poems are unsettling: the first leaves unanswered the question of whether the speaker himself is a ghost; the second focuses on the way that the occupants of the cabin are *talking in the dark* (34), and this sense of their lack of direction could be interpreted as the *inner haze* (42) of the last line. Questions about identity and about knowing the direction of one's life occur all the time in Frost's poetry, so it is revealing to find them appearing in this pair of poems written at opposite ends of his career. The setting of the first poem could suggest a deserted farmstead (the *grapevines* [6], the *mowing field* [7] and the *orchard* [8]). Is your interpretation of the poem affected by knowing that the farm in which Frost and his wife lived for several years from 1901 onwards was so isolated that they had hardly any visitors at all? Is it necessary, helpful or irrelevant in reading *A Cabin in the Clearing* to understand something of the tension between the native American Indians and the pioneers who settled in territory that had belonged to the Indians?

Frost's Titles

Always begin with the titles: they may supply valuable clues either about the meanings of individual poems or about the poet's wider themes and concerns. The titles of Frost's collections of poems, from *North of Boston* (1914) right through to *In the Clearing* (1962), nearly all focus on places or on features of the landscape. References to Boston and New Hampshire locate the poetry very clearly in a particular geographical setting: the area where Frost lived and farmed as a young man, and in which he felt most at home throughout his life.

Usually, the titles of Frost's poems suggest the key image or theme of the poem (*Mending Wall*, p.9, *Desert Places*, p.48); sometimes they almost seem like labels attached to photographs in an album (*The Black Cottage*, p.21, *Stopping by Woods on a Snowy Evening*, p.41). Occasionally, there is a note of old-fashioned formality about a title (*To E.T.*, p.40, *For John F. Kennedy His Inauguration*, p.62) or of sheer oddity (*The Oven*

Bird, p.29, *An Unstamped Letter in Our Rural Letter Box*, p.56) which may affect the reader's approach to the poem that follows.

It is also important to review the title after reading the poem: for instance, one of Frost's most familiar poems is entitled *The Road Not Taken* (p.28) but really the poem's subject is the road that *was* taken, not the one that wasn't. *The Most of It* (p.52) is an example of an ambiguous title, which forces the reader to ask what Frost is really referring to; does the final comment of the poem (*and that was all* [20]) imply that the experience described in the poem was ultimately significant or trivial?

Activity

Look at all the titles in the Contents, and first list the poems whose titles suggest life and work on a farm or in a rural environment. Now look up these poems: what is the function of the title – is it simply there to introduce the subject, or does it obliquely or directly comment on it? Are there other poems which refer to life on a farm but whose titles do not signal this?

Discussion

While some of the titles seem simply related to rural life and work (*Mowing, Mending Wall, After Apple-Picking*, etc.) they actually direct the reader to the significant discovery or conclusion reached as a result of the action described in the title and in the poem. Thus, *Mowing* (p.5) not only describes the action and the sound of mowing with a scythe, it asserts that the action of mowing is its own fulfilment: *The fact is the sweetest dream that labor knows* (13). Judith Oster (*Toward Robert Frost: The Reader and The Poet*, p.64) says of this poem:

> Frost seems at times to want to retreat from metaphor, to deny it altogether with a sort of antimetaphor poem that makes no claim to mean anything other than the experience it presents – mowing and leaving the mown grass to 'make' itself into hay. In *Mowing* fact and poetic act are so inextricably fused that we do not draw lines of analogy. The fact *is*, seemingly with no intervening imagination or act of mind.

At the same time, the poem implies that all work can be its own reward, and that what may at first seem only destructive (cutting grass) is actually creative (making hay). In this sense, mowing can be seen as an image of all creative work (the work of a poet included) – the poem's meaning moving outwards from the particular to the universal.

A poem such as *The Ax-Helve* (p.36) is set specifically in the type of small, remote farming community with which Frost was familiar. The serpentine shape of the axe handle (the helve) reflects the tortuous direction of the poem and the stages by which the speaker and his neighbour establish a relationship. It also represents the instinctive knowledge and expertise that the speaker's neighbour, Baptiste, has and which is so important to his self-esteem. On the other hand, for the speaker himself the ax-helve is a very ambiguous symbol:

> But now he brushed the shavings from his knee
> And stood the ax there on its horse's hoof,
> Erect, but not without its waves, as when
> The snake stood up for evil in the Garden—[i.e. of Eden] (94–97)

Here, therefore, the title focuses attention not just on the central image of the poem, it defines the key themes of the poem. It also embodies the moral disquiet felt by the speaker when he reflects on the way the ax-helve has been used by his neighbour to establish his superiority. Very often, when reading Frost, we are forced to reassess the significance of his titles; and, in so doing, we realize the complexity and richness of his poetry.

Frost's Life

Early Life and Farming

Frost was born in 1874 on the west coast of the United States, in San Francisco, where his father was a moderately successful journalist and editor, public speaker and local politician. His parents are summed up by William H. Pritchard in the following terms:

> William Frost was preoccupied, severe, and short-tempered, a man who believed in discipline and in wasting no words; a man of action rather than contemplation; a drinker, with the conviction that the real things of this world were public ones – the wars of politics and journalism. His inadequacies, his failures even, were ... brought out more sharply through their juxtaposition with a woman surely as

> spiritual, idealistic and powerful in her moral goodness as could be imagined. (*Frost: A Literary Life Reconsidered*, p. 31)

Frost's mother, Isabelle, was extremely protective of her son, who was born after his parents had been married a year. For part of his childhood she taught him herself at home, and when her husband died in 1885 she moved across the continent to start a new life with her children. She became a teacher, a career which Frost himself was to follow for much of his life.

It was the move to the eastern states of America, to New England – that area consisting of Massachusetts, Vermont and New Hampshire – which had a defining impact on Robert Frost both as an individual and as a poet. Here he lived for most of the rest of his life, always returning if he moved temporarily somewhere else. Although he was an extremely restless man (it has been estimated that he lived in over seventy houses during his adult life), the rural landscape of New England shaped his life and dominated his imagination. Curiously for someone who was to spend much of his career teaching in universities and colleges, Frost twice began courses at university (Dartmouth and Harvard) and twice failed to complete them. On both occasions it was to his familiar environment that he returned, and it was farming and teaching in New England that occupied most of the first forty years of his life.

Although he liked to give the impression of being an unsophisticated, rather conservative, reclusive countryman, Frost was very well read in English literature and the classics. The influence of Shakespeare, Milton and the Romantic poets (Wordsworth, Keats and Shelley in particular) can often be traced in his own writing. Unlike many aspiring writers, however, he made very little effort at first to meet other poets and novelists. The apprenticeship he served in poetry was a long and solitary one. In 1899, already married and with a young family, he took up chicken farming. Two years later, an uncle generously purchased a small-holding in Derry, New Hampshire, for Frost to farm; here he lived a life of apparently remarkable isolation, where visitors were few and far between, and where he wrote poetry in the evenings after the day's work was completed and the children had been put to bed. Frost had met his wife, Elinor, in 1891 while they were both still at school. They married four years later in 1895. Although she was the inspiration behind, or the subject of, some of his most memorable poems, their relationship was not a straightforward one. Their first son, Elliott, died in 1900 at the age of three and the tensions this aroused between them provoked one of Frost's finest poems, *Home Burial* (p.17). (See Notes, p.73)

Activity

Read *The Pasture* (p.1), *Home Burial* (p.17) and *The Death of the Hired Man* (p.11). How far does a knowledge of the biographical background to the poems deepen or detract from your appreciation of them as poetry?

Discussion

The Pasture is one of Frost's shortest poems; *Home Burial* and *The Death of the Hired Man* are two of his longest. In *The Pasture* the speaker identifies himself as someone who has work to do on the farm, and who invites the listener to come with him to share the experience or at least keep him company. The repeated line *I shan't be gone long. – You come too* (4, 8) suggests either a reluctance on the part of the speaker to be separated from his companion or an awareness that she (if it is Elinor) does not want to be left alone again, even for a short time. The lure of the outdoors (the pasture spring which needs clearing, the calf that is so new it is still unsteady on its feet) may be strong; alternatively, the demands of the farm may be putting a strain on their relationship, reflected by his first saying he will not be away long and then in the next breath inviting her to join him if she resents his absence.

Home Burial shows a relationship at breaking point. The frustration of the husband at being rejected by his wife (*Let me into your grief* [59]) and the wife's horrified recollection of the relish with which her husband had dug the little grave for their child (*Making the gravel leap and leap in air* [75]) makes this one of the bleakest of Frost's poems. In what sense, if any, can we also call it 'honest'? Can we be sure it is intended to be an accurate record of the relationship between Robert and Elinor Frost? After all, the woman in the poem is specifically identified as *Amy* (39). If the situation and the dialogue sound convincingly authentic, does it have to be because of what we know of Frost's marriage at this time? The careful description of every movement made by the couple as the wife edges closer and closer to the door, the ebb and flow of their voices as the argument gathers momentum, and the impotent final cry of the man (*I'll follow and bring you back by force. I will!—* [116]) – all of these reflect the poetic as well as the dramatic control exercised by Frost over his material. It is this control which gives the poem its remarkable force.

The Death of the Hired Man seems to depict a relationship halfway between those presented in the other two poems. Although the couple, Warren and Mary, disagree about the motives of Silas, the hired man, in returning to their farm, and although Warren sneers at the word *Home* (113), it is clear that there is a rapport between the two; and Frost's more relaxed control of the dialogue means that there is less of the strain so tangible in *Home Burial.* By the end of the poem, Silas's death helps to (re-)unite the couple:

> Warren returned – too soon, it seemed to her—
> Slipped to her side, caught up her hand and waited. (164–165)

The farm is of course one of the most common and most significant settings in Frost's poetry. Although he nearly always owned a farm, he was rarely content to devote himself exclusively to farming: five years after he took on the farm at Derry in 1901, he began teaching at a local school, Pinkerton Academy, leaving Elinor to do much of the daily work on the farm. In later life, he kept a farm as a kind of summer retreat and expected his children to help with the day-to-day management of it. It might be more accurate to think of Frost as a countryman rather than specifically as a farmer: his poems after all deal particularly with the life of the country and the people and creatures who inhabit it – the neighbour Baptiste in *The Ax-Helve* (p.36); the boy who has the fatal accident with the saw in *'Out, Out–'* (p.34); the oven bird whose midsummer singing warns of the coming of autumn; the buck and doe in *Two Look at Two* (p.42) who encounter the lovers (possibly Frost and Elinor, but does a reading of the poem gain anything from this possibility?).

Frost's imagination was most richly fed by the New England countryside. It was in this environment that his private moods and obsessions could best be explored and transformed through poetry. In turn, this particular landscape itself is universalized through the act of reading, so that an appreciation of poems such as *Birches* (p.30), *Stopping by Woods on a Snowy Evening* (p.41) or *The Most of It* (p.52) does not depend on our being able to recognize the settings as authentically New England. The reader is invited (*You come too*, *The Pasture*, p.1, lines 4 and 8) to inhabit the landscape imaginatively, just as Frost himself inhabited it physically.

Frost, England and Edward Thomas

In the summer of 1912, Frost, with his wife and four children, sailed to England hoping to find a home near London which would enable him to write and to make literary contacts helpful to his still-intended career as a poet. At this time, he was thirty-eight and had not yet published a collection of his poems.

They rented a house in Beaconsfield, some way north west of London, and Frost was able to make the acquaintance of several of the younger poets and critics of the day. The first volume of a new poetry anthology, *Georgian Poetry*, was published in 1912; the contributors to this included Rupert Brooke, Wilfred Gibson and Lascelles Abercrombie – a nucleus of young writers who came to be known as the Georgian poets. Frost was interested in their work, and got to meet them. He particularly liked W.W. Gibson and the critic and reviewer, Edward Thomas (who, although not strictly speaking a Georgian poet, was close to several of the writers in the group). For the first time Frost began to feel at home with other writers. In 1913 Frost published *A Boy's Will*, and *North of Boston* in the following year, to favourable reviews. The paragraphs below come from reviews by Edward Thomas, who was at this time one of the most influential and perceptive critics of modern poetry in England.

> This [*North of Boston*] is one of the most revolutionary books of modern times, but one of the quietest and least aggressive. It speaks, and it is poetry. It consists of fifteen poems, from fifty to three hundred lines long, depicting scenes from life, chiefly in the country, in New Hampshire... These poems are revolutionary because they lack the exaggeration of rhetoric, and even at first sight appear to lack the poetic intensity of which rhetoric is an imitation. Their language is free from the poetical words and forms that are the chief material of secondary poets...
>
> The language ranges from a never vulgar colloquialism to brief moments of heightened and intense simplicity. There are moments when the plain language and lack of violence make the unaffected

> verses look like prose, except that the sentences, if spoken aloud, are most felicitously true in rhythm to the emotion. (Edna Longley, (ed.), *A Language Not To Be Betrayed: Selected Prose of Edward Thomas*, pp.125 ff.)

Thomas was four years younger than Frost. They had first been introduced to each other in a London tea-shop; the friendship that was to develop between them (broken only by Thomas's death in the trenches in 1917) was of vital importance to the lives and careers of both men.

In 1914, Frost moved with his family to the remote village of Dymock in Gloucestershire where Gibson and Abercrombie had gathered together a group of friends who included (from time to time) Rupert Brooke and Edward Thomas and his family. Here the friendship between Frost and Thomas grew quickly as they discussed poetry and their ideas about poetic speech. (See Edward Thomas's poem *The Sun Used to Shine*, Appendix, p.140.) Frost claimed that: *Words are only valuable in writing as they serve to indicate particular sentence sounds*. (Letter to John Bartlett, 4 July 1913.) He believed that in poetry the rhythm of a sentence was as important as the meaning of the words, and that the ebb and flow of the speaking voice talking normally (not trying to create special poetic effects) was what a poet had to recreate. Thomas agreed, and when with Frost's encouragement he started to write poetry himself, the very naturalistic way in which he made blank verse sound like speech, while retaining the rhythmic power of poetry, was a major feature of his verse.

Activity

Compare the style and tone of Thomas's poems *Wind and Mist* (p.146) and *As the Team's Head-Brass* (p.142) with Frost's *Mending Wall* (p.9) and *Home Burial* (p.17). Does reading these poems (aloud if possible) help you to understand why, for Frost, the spoken sentence was such an important rhythmic unit in a poem?

Discussion

Dialogue is one of the most important elements of many Frost poems. *Mending Wall* and *Home Burial* are written in blank verse (unrhymed lines of ten syllables with the stress usually falling on the second, fourth, sixth, eighth and tenth syllables). However, the dialogue they contain is written in sentences, not lines; frequently a sentence will begin or end at the middle of a line or overrun the line ending. This may create a tension between the rhythm of the spoken sentence and the metre (regular rhythmical pattern) of the blank verse. When reading the poems by Thomas and Frost are you more aware of the metre of the verse or the cadence (stress) of the spoken sentence?

In *As the Team's Head-Brass*, the iambic rhythm of the line is balanced by the rhythms of the dialogue between the speaker (the poet himself?) out walking in the country and the ploughman whom he meets. Their conversation is itself punctuated by the gaps during which the ploughman is following the team of horses from one side of the field to the other. Thus, the ebb and flow of their discussion about the war is further reflected in the staccato question and answer form of their dialogue.

Frost once wrote: *Remember that the sentence sound often says more than the words. It may even as in irony convey a meaning opposite to the words.* (Lawrance Thompson and R.H. Winnick, *Robert Frost*, p.172) Thus, the repeated and emphatic statement in *Mending Wall – Good fences make good neighbors* (27) – has been taken by many readers as the sort of homely proverb of which Frost himself might approve; others find it irritatingly glib and deduce from this that Frost distrusts such a confident assertion that people are better kept apart. Does the tone of *Mending Wall* give you a clear sense of Frost's attitude?

Frost and Thomas had much in common besides their interest in poetry. They had similar temperaments (both could be moody, difficult to live with, yet aware of how much they hurt the people they loved – see, for example, the Notes to *Home Burial* (p.73), for comments on Frost's relationship with his wife) and their careers had not really developed as either had hoped. Thomas had had to devote nearly all his time to reviewing and freelance work to make enough money to support his family. What they did, as they talked, was to encourage each other to believe in themselves as poets. Frost showed Thomas that the prose he was already writing had the rhythms of verse and could be reshaped to become poetry. After Thomas's death, Frost wrote:

> We were greater friends than almost any two practising the same art. I don't mean that we did nothing for each other. As I have said we encouraged each other in our adventurous ways. Beyond that

> anything we did was very practical. He gave me standing as a poet – he more than anyone else... I dragged him out from under the heap of his own work in prose he was buried alive under... All he had to do was to put his poetry in a form that declared itself. (From a letter to Edward Garnett)

After the outbreak of the First World War, Frost and his family returned to America. Thomas had at first been tempted to go with them; in the end, his son accompanied the Frosts and Thomas joined the army. He and Frost never met again, although they continued to correspond until Thomas was killed. Frost later summed up their closeness in the poem *To E.T.* (p.40):

> I meant, you meant, that nothing should remain
> Unsaid between us, brother... (9–10)

Their friendship, although cut short by war, proved to be one of the most significant events in the lives of both men.

Activity

Compare *The Sun Used to Shine* (p.140) by Edward Thomas with Frost's *To E.T.* (p.40).

Discussion

The Sun Used to Shine, as its title implies, looks back to a time when *we two* (1) – Thomas and Frost – were able to take such pleasure in each other's company that it was hard to focus on the reality of the war that would separate them for ever. The careful organization of the poem into quatrains (four-line stanzas) does not disrupt the flow of the long sentences which make up the poem: only one stanza, apart from the final verse, ends with a punctuation mark – and that is a semi-colon rather than a full stop. This flow of recollection culminates in the moving final sentence which is spread across three stanzas. The poet both anticipates his own death (*Everything/To faintness like those rumours fades* [22–23]) and looks forward to a future in which others may share the same delight in friendship that he has shared with Frost:

> And other men through other flowers
> In those fields under the same moon
> Go talking and have easy hours. (30–32)

Frost's poem is more formal, as its title and diction indicate: *slumbered* (1), *brother* (10), *Victory* (12), *the shell's embrace* (13) suggest the rhetoric of war as a romantic and heroic activity. In this sense the poem is intended as a tribute to the memory of Thomas, *First soldier, and then poet* (7). However, in the final stanza, Frost asks poignantly how the war can be over, if he cannot share its conclusion with his friend?

> How over, though, for even me who knew
> The foe thrust back unsafe beyond the Rhine,
> If I was not to speak of it to you
> And see you pleased once more with words of mine? (17–20)

The greater loss, therefore, is Frost's and this becomes the theme of the poem. It is significant that although it might have been thought that Thomas, encouraged by their friendship to turn himself into a poet, owed more to Frost than vice versa, it is the survivor (Frost), who still longs for the dead friend's approval of his work. The sense of loss, and the feeling of incompleteness as a consequence of loss, is a key and recurrent theme in Frost's poetry.

The Growth of Frost's Reputation

North of Boston, published in 1914, made Frost's reputation, and for some critics it remains his finest collection of poems. However, his growing popular appeal was reflected by the increasing sales he achieved with each new volume and his critical status was ensured by his four Pulitzer Prizes (the leading American literary awards), the first of which he won in 1924. By now he had begun to earn a regular income from the royalties of his books and from teaching. This teaching was always fairly informal, and allowed him time to write and to travel widely across the United States, giving poetry readings and lectures. He developed a style of relaxed, rather homespun speaking and reading which attracted large and enthusiastic audiences. In a time before television and radio, these performances gave him a vital contact with the reading public and kept his reputation as a poet alive.

For his part, Frost relished his role as a performer, and continued to give lectures and readings to the end of his life. The sheer size of his audiences (8,500 was the largest he addressed in person, excluding his appearance at President Kennedy's inauguration in January 1961) led some critics and other poets to ask whether his popularity had not become greater than his poetry deserved.

One reason for Frost's appeal was that his poetry was thought to be accessible and not complex. During the 1920s works by influential modernist authors such as T. S. Eliot and James Joyce seemed to herald a new era of writing which was too obscure and intellectual to appeal to the common reader. In poetry, particularly, works such as *The Waste Land* (1922) challenged so many preconceptions about what poetry was, or ought to be, that all contemporary poets were judged to be either traditionalists or modernists. Frost himself was happy to attack obscurity in poetry, though this camouflaged the fact that much of his writing of this period was itself very far from straightforward and that his own technique sometimes involved echoing or directly borrowing from other poets (a tendency he criticized in the poetry of Ezra Pound and T. S. Eliot) to create his own poem.

A poem as apparently uncomplex as *Stopping by Woods on a Snowy Evening* (p.41) involves such borrowing. Jeffrey Meyers (*Robert Frost, A Biography*, p.180) has pointed out that the line *He gives his harness bells a shake* (9) echoes a line from *The Rover* by Sir Walter Scott: *He gave the bridle-reins a shake*. Similarly, *The woods are lovely, dark, and deep* (13) is very close to *Our bed is lovely, dark and sweet*, which comes from *The Phantom Wooer* by Thomas Lovell Beddoes. Both poems were well-known to Frost; either he had unconsciously absorbed the lines and adapted them in creating his own poem, or he was deliberately drawing on his own poetic memory to give his poem additional resonance. Part of the effect of *Stopping by Woods on a Snowy Evening* (p.41) comes precisely from the fact that it sounds so familiar, like a poem one has somehow always known.

How poets absorb, remember and adapt each other's work is a fascinating and important question. One way of measuring the growth of a poet's reputation and influence is to judge his impact on other writers, particularly his contemporaries or those of a younger generation. Frost was older than writers such as Ezra Pound and T. S. Eliot, but because he had begun his career as a published poet so much later in his life than they did (he was forty when *North of Boston*, his second book, made his name both in Britain and in America; Eliot was twenty-nine when *Prufrock and Other Observations* was published in 1917) he seemed to belong to their generation.

Frost certainly drew on Eliot – compare, for instance, Frost's *Acquainted with the Night* (p.46) with Eliot's *Preludes* – but neither Eliot nor Pound (though they admired his writing) drew on Frost. On the other hand, younger poets have been open in acknowledging their debt to Frost: Seamus Heaney, the Nobel Prize-winning Irish poet, has described *Stopping by Woods on a Snowy Evening* (p.41) as his favourite poem and W.H. Auden, another poet who was quick to admire the writing of Frost, clearly echoes the repeated last line of Frost's poem in the final stanza of his own poem, *Their Lonely Betters*:

> We, too, make noises when we laugh or weep:
> Words are for those with promises to keep.

Another factor behind the popular success of Frost's poetry was that the stoical, individualistic philosophy he seemed to be spelling out appealed strongly to a public trying to come to terms with the major historic events of the Twentieth century. Frost had been in England when the First World War broke out in 1914; he lived and wrote his way through the industrialization of the United States, the Wall Street Crash of 1929 and the Depression, the Second World War and the Holocaust; he actually went to Russia as a cultural ambassador and met the Soviet leader, Khrushchev, at the height of the Cold War, just before the Cuban Missile Crisis of 1962. These events, however, rarely featured (even obliquely) in his poetry; instead, Frost continued to draw on the rural and remote life of his early years as if it existed in a kind of continuum unaffected by the traumas of the world outside. In doing so he helped reinforce a sense of an American identity (modest but proud to be self-reliant, undaunted, close to nature) which was an attractive 'idealistic' alternative to the urban, anxious, materialistic society most Americans actually inhabited. Frost deliberately drew on the historical tradition of the New England settlers of the Sixteenth and Seventeenth centuries – puritan, self-sufficient, pioneering, libertarian in their belief in the freedom of the individual and their opposition to slavery. In essence, these were the values enshrined in the Declaration of Independence. It could be suggested, therefore, that Frost in his poetry offers an important redefinition of the American Dream.

Activity

Read *The Black Cottage* (1914) (p.21) and *For John F. Kennedy His Inauguration* (1962) (p.62). (The dates in brackets refer to the date of publication in a volume of poems.) How does the first poem reflect Frost's attitude to American history and values? Do these attitudes and values re-appear in the later poem?

Discussion

The old lady who used to occupy the Black Cottage comes to represent the traditional values of white America which the speaker, the minister who is accompanying the narrator and whose monologue in fact occupies most of the poem, strongly endorses. Both her husband and her two sons had been killed in the American Civil War; the minister says that for her, it was not the Union of the States or the abolition of slavery which made the sacrifice worthwhile, but:

> Her giving somehow touched the principle
> That all men are created free and equal. (60–61)

The old lady embodies, for the minister, values and views which have become unfashionable but which should still be cherished. He admits that she had scarcely met any black or coloured people; but since they were all children of God (*the same hand working in the same stuff* [78]) to her, questions of race and colour were unimportant. In the end, the minister argues, truth is what people choose to believe and there is no such thing as a new idea:

> For, dear me, why abandon a belief
> Merely because it ceases to be true.
> Cling to it long enough, and not a doubt
> It will turn true again, for so it goes.
> Most of the change we think we see in life
> Is due to truths being in and out of favor. (105–110)

This could be interpreted as Frost's own philosophy, a touching restatement of the belief that there is nothing new under the sun and that progress is an illusion. On the other hand, Frost is careful *not* to say that he personally subscribes to this philosophy; he chooses instead to filter the views of the old lady through the mouth of the minister, whose words are then reported – without comment – by the narrator, *I*. The cottage in which the old lady had lived is deserted but untouched; the furniture in it is as old-fashioned as the views she had held.

The poem has a wonderfully unexpected and ambiguous ending. The minister, who has been fantasizing about being monarch of a desert island devoted to preserving ideas that have gone out of fashion, breaks off in mid-sentence:

> 'There are bees in this wall.' He struck the clapboards,
> Fierce heads looked out; small bodies pivoted.
> We rose to go. Sunset blazed on the windows. (125–127)

After the minister's rambling monologue, these abrupt statements have a peculiar force, emphasized by the strong caesura (a pause or break, created here by the punctuation) in each line. The bees seem like the guardians of the house, protecting the old lady's memory and her values. Meanwhile, though the cottage is black and impenetrable (the narrator and the minister cannot enter, they simply peer through the window and then sit on the doorstep) the sunset lights up the glass of the window panes. Is this the light of the present illuminating the darkness of the past? Or might the fact that it is sunset suggest that the wisdom of the present is itself fading? Is it significant that the sunset blazes *on* (127), not through, the windows? The imagery of the final lines could suggest all these possibilities. The minister claims:

> Whatever else the Civil War was for,
> It wasn't just to keep the States together,
> Nor just to free the slaves, though it did both.
> She wouldn't have believed those ends enough
> To have given outright for them all she gave. (55–59)

That last line foreshadows the title of the poem Frost read at President Kennedy's inauguration. The subtitle of *For John F. Kennedy His Inauguration* is *Gift outright of 'The Gift Outright'* (*With some preliminary history in rhyme*). The word *outright* emphasizes that the commitment of the American people to the land which *was ours before we were the land's* (1) is unconditional. In the election of the young, charismatic Kennedy, Frost saw *The glory of a next Augustan age* (71) approaching; characteristically, though, he also warned that *There is a call to life a little sterner* (66). For some, this poem was a rousing and appropriate assertion of American identity and confidence; to others, it was empty rhetoric whose talk of *A golden age of poetry and power* (76) was unconvincing and dishonest.

Frost's Imagery and Themes

This section will approach Frost's poetry through a study of some of his key imagery and themes, and will suggest how a reading of one poem can throw light on the interpretation of others. There is a remarkable consistency in his writing, but when he returns to a previously used image or theme, Frost never simply recycles it: a fresh encounter leads to new perspectives.

Leaves, logs, trees, woods, forests: all of these feature repeatedly in Frost's poetry and the idea of the cycle of nature and of life generally is a theme to which he keeps returning. At the start of his career a poem such as *In Hardwood Groves* (p.8), (1913) is a meditation on the progress of leaves: first they are found *giving shade above* (2), then they fall and *fit the earth like a leather glove* (4). These leaves must rot into the ground before the next generation of leaves can replace them: *They must go down* (7), Frost repeats and then stresses, *They must be pierced by flowers and put/Beneath the feet of dancing flowers* (9–10). This cycle is part of the human condition and the poem's final point is that such a cycle is inescapable:

> However it is in some other world
> I know that this is the way in ours. (11–12)

The sheer inevitability is something which has to be accepted, and Frost's fatalism can be heard clearly in these lines. In a later poem, A *Leaf-Treader* (p.49), (1936) the image of the leaves is given a more sinister colouring. The poem begins with the speaker exhausted after a day spent gathering leaves and treading them into a compost heap; in so doing, he has (as it were) been gathering up the evidence of the past twelve months: *I have safely trodden underfoot the leaves of another year* (4). If there is a sense of relief here, it is quickly undermined in the second stanza:

> All summer long they were overhead, more lifted up than I.
> To come to their final place in earth they had to pass me by.
> All summer long I thought I heard them threatening under
> their breath.
> And when they came it seemed with a will to carry me with
> them to death. (5–8)

Although the tone here is a good deal more weary and foreboding (compare the length and rhythm of these lines with the metrical form of *In Hardwood Groves*, p.8) the basic idea is the same: the leaves begin overhead and end up underfoot. Here, though, the speaker half feels that the leaves want to carry him with them into the ground. In the final stanza he acknowledges that the idea is attractive: *They spoke to the fugitive in my heart* (9). The temptation to succumb, to go into the dark, (to commit suicide?) is as real here as it is in *Stopping by Woods on a Snowy Evening* (p.41) where *The woods are lovely, dark, and deep* (13) but the speaker accepts his responsibility to continue living:

But I have promises to keep,
And miles to go before I sleep, (14–15)

A *Leaf-Treader* (p.49) similarly ends with an acknowledgement that he must keep going. He has to tell himself to continue making the effort to live: *Now up, my knee, to keep on top of another year of snow.* (12). For Frost, leaves and snow perform the same function: they cover the landscape in autumn and winter and emphasize to the poet his sense of his own impermanence. *Desert Places* (p.48) contains a candid portrait (a self-portrait?) of an individual struggling to come to terms with this sense. Snow and darkness fall together, *fast, oh, fast* (1): the fields are almost covered, the surrounding woods have already been cloaked by it and the animals are *smothered in their lairs* (6). By contrast, the speaker feels that the snow has ignored him:

I am too absent-spirited to count;
The loneliness includes me unawares. (7–8)

But he is no stranger to this experience. Though he knows the loneliness will get worse he is not frightened by the *blanker whiteness of benighted snow* (11) and claims that the woods *cannot scare me with their empty spaces* (13):

I have it in me so much nearer home
To scare myself with my own desert places. (15–16)

After reading this final couplet we cannot avoid re-reading all the poems in which Frost presents us with an empty landscape or one obliterated by leaves or snow: are they all images of his own, inner *desert places* (16)?

Activity

Read *Tree at My Window* (p.45). How does this poem compare with *Desert Places* (p.48)?

Discussion

A similar awareness to that expressed in *Desert Places* can be found in *Tree at My Window* where the speaker, awake in bed at night, compares the storm-tossed tree outside with himself:

> But, tree, I have seen you taken and tossed,
> And if you have seen me when I slept,
> You have seen me when I was taken and swept
> And all but lost. (9–12)

The exterior world of storm mirrors his own personal turmoil and this agitation is expressed poetically and dramatically by the alliteration in the stanza, by the insistent repetition within three and a half lines (*you...you...You; I...I...I; have seen...have seen...have seen* [9–12]) and by the emphatic rhymes (*tossed/lost* [9, 12]; *slept/swept* [10, 11]). Finally, Frost makes the point explicitly:

> That day she put our heads together,
> Fate had her imagination about her,
> Your head so much concerned with outer,
> Mine with inner, weather. (13–16)

However, the poem can be approached rather differently. Judith Oster (*Toward Robert Frost: The Reader and The Poet*, p.146) sees the relationship between the speaker inside the house and the tree at his window in these terms:

> ... the tree, in one sense, externalizes and represents what is of major concern to the poet – dreaming, articulating, and being tossed by some force more powerful than tree or person... The tree has height, but the man has depth. Likewise the tree's concern is only with outer weather, for only man can suffer 'the tempest in [the] mind.' The tree at the window becomes the window tree – part of the man's home, though safely outside it.

Frost's Philosophy

Of all the trees that Frost names in his poems, it is the birch that features most often, and his poem *Birches* (p.30) offers one of his most important statements about man's place, not just in the natural world but in the universe. Whether or not the description of the child climbing the birch tree –

> Some boy too far from town to learn baseball,
> Whose only play was what he found himself (25–26)

– is autobiographical, the whole poem has the ring of authenticity. Frost uses extraordinary detail to evoke the sights and sounds in a wood when the *enamel* (9) of ice which has coated the birch tree in the early morning begins to fall onto the snow crust below. Such sights may be completely unfamiliar to the reader but Frost, by his carefully placed, casual familiarity (*Often you must have seen them* [5]) conveys a sense of something that is both known and unknown at the same time. His account of the skill with which the child is able to judge the exact moment at which to swing the tree climaxes on a note of nostalgia and achievement: *So was I once myself a swinger of birches* (41). The poem could well have ended at this point, but in the ensuing lines Frost explores the image of the child climbing *Toward heaven* (56) and then returning to earth as a metaphor for human aspiration and limitation. The speaker remarks that the urge to become once again *a swinger of birches* (41) comes over him when

> ... life is too much like a pathless wood
> Where your face burns and tickles with the cobwebs
> Broken across it, and one eye is weeping
> From a twig's having lashed across it open. (44–47)

Notice here how Frost combines the symbolic *pathless wood* (44) with the vivid description of what it feels like actually to walk through such a wood; as in *Tree at My Window* (p.45), the inner and the outer worlds of experience are reflections of each other. The poem ends with an acceptance of the value of life (*Earth's the right place for love:/I don't know where it's likely to go better* [52–53]) but also with a wish to be able to escape temporarily from it. The last lines sum up the poem with two characteristically low-key statements:

> That would be good both going and coming back.
> One could do worse than be a swinger of birches. (58–59)

One could do worse (59) may sound a rather grudging endorsement, but it combines a possible hint of mischief in its use of litotes (ironical understatement) with a recognition that life offers few opportunities for safe escape. This sense that all life is provisional and that its continuance must not be taken for granted is a theme that occurs very frequently in Frost's poetry; it can be seen as a key element of his philosophy. It is treated forcefully in poems like '*Out, Out–*' (p.34) and *The Draft Horse* (p.66) and more indirectly but hauntingly in *The Oven Bird* (p.29) where, in the height of summer and in the depth of a wood, a warbler warns that autumn is coming (*He says that leaves are old* [4]). Frost's conclusion to this poem is ambiguous:

> The question that he frames in all but words
> Is what to make of a diminished thing. (13–14)

What to make of... (14); does he mean 'How can one make sense of'? or 'How far can one go on enjoying and celebrating something (summer; life itself), knowing it is shortly to end?'

Activity

Read *Old Man* by Edward Thomas (Appendix, p.144) and compare its imagery and themes with those of *Birches* (p.30). How far, and in what ways, does your reading of the one illuminate your reading of the other?

Discussion

Both poems contain very precise descriptions of things observed in nature; both also involve memory and childhood, though in *Old Man* the speaker's memories are less defined than in *Birches*: he knows that the strong-smelling herb holds the key, but he cannot unlock the memory and the poem ends with the terrifying list of seven negatives followed by the final image – *Only an avenue, dark, nameless, without end.* (39). In this respect, Thomas's poem is the bleaker of the two and shows up (by contrast) the element of conditional optimism that Frost exposes in the final lines of *Birches*. There are options and possibilities open to Frost –

> I'd like to get away from earth awhile
> And then come back to it and begin over. (48–49)

– which are not open to Thomas in *Old Man*. The abiding tone of each poem is contained within its last line. Nevertheless, in Thomas's poem the speaker is able to wish a future in which the child will remember where first she smelt the herb; the pessimism of the ending arises from a personal failure. Thomas's acute sense of 'lost-ness' and loneliness is perhaps closer to the sense of 'blankness' and 'absent-spiritedness' which Frost expresses so movingly in *Desert Places* (p.48).

One of the most striking similarities between *Old Man* and *Birches* is the way in which both poets present the reader with images of nature that are acutely realized in themselves (i.e. vividly and precisely described as the things themselves – the scent of the herb, the bending of the birches) and yet which are also powerful metaphors; they appeal to the reader's sense of reality and to his or her imagination. In this, Judith Oster *(Toward Robert Frost: The Reader and The Poet*, p. 61) suggests, *we see Frost's typical 'between-ness', the pull of contraries that keeps him in balance.*

Frost's Voices

The Narrative Voice

Some years after the end of the First World War, W.W. Gibson published a recollection of the Dymock poets in a poem called *The Golden Room*:

> Do you remember that still summer evening
> When, in the cosy cream-washed living-room
> Of the Old Nailshop, we all talked and laughed—
> Our neighbours from The Gallows, Catherine
> And Lascelles Abercrombie; Rupert Brooke;
> Eleanor and Robert Frost, living a while
> At Little Iddens, who'd brought over with them
> Helen and Edward Thomas? In the lamplight
> We talked and laughed; but, for the most part, listened
> While Robert Frost kept on and on and on,
> In his slow New England fashion, for our delight,
> Holding us with shrewd turns and racy quips,
> And the rare twinkle of his grave blue eyes? (1–13)

This memory of Robert Frost dominating the conversation among friends is an apt starting point for a discussion of his 'voice'. A poet's 'voice' is the distinctive (and probably unique) style and tone of speaking through poetry which a reader can recognize and identify. It becomes an essential part of the texture of a poem; to be familiar with a poet's 'voice' is to be better able to pick up the underlying tone of his or her writing, registering hints of irony, for example, or sudden shifts of mood which might not be so quickly apparent to someone unfamiliar with this voice. Gibson's account of Frost's *slow New England fashion* (11) of speaking evokes at once the slow, relaxed voices of the speakers in many of Frost's own poems. These voices are not only heard in the poems which consist largely of dialogue, but in those where the narrator takes his time to lead the reader to the point of the story.

Activity

Contrast *An Encounter* (p.33) with *'Out, Out—'* (p.34) to examine the differences and similarities in the way the 'voice' in each is established and used.

Discussion

Both poems tell a story, so it is appropriate to talk of the speaker as a narrator. *An Encounter* has a first-person narrator recounting an incident that occurred to him; *'Out, Out—'* also has a first-person narrator, but he only makes one direct intervention (*Call it a day, I wish they might have said* [10]) in what is otherwise a terse account of an accident involving people with whom he may have had no personal connection at all.

In the first poem, the speaker takes precisely half of its twenty-four lines to reach the encounter which is the subject of the poem: his meeting with what turns out – half-comically – to be a telegraph pole. It is a hot day, and the narrator's slow delivery, as he winds his way towards the encounter, is matched by the serpentine rhyme scheme of the poem. (Work out the rhyme scheme for yourself: can you see any pattern in it or is it simply random?) Words such as *slowly* (2), *weary* (6), *overheated* (6), *paused* (8), *rested* (8) establish the mood, and the very ordinariness of the diction here leaves the reader unprepared for the dramatic central image of the poem:

... a resurrected tree,
A tree that had been down and raised again—
A barkless specter. (12–14)

By contrast, the second poem begins in the middle of the story (*The buzz saw snarled and rattled in the yard* [1]) but that opening statement is shortly taken up again and repeated twice to emphasize the endlessness of the day's activity:

And the saw snarled and rattled, snarled and rattled,
As it ran light, or had to bear a load.
And nothing happened... (7–9)

The refusal to hurry the narrative persists even at the moment of the accident, as if the speaker is deliberately refusing to be sensationalist. Frost's *slow New England fashion* (see *The Golden Room*, p.121) is nowhere more evident than here:

... At the word, the saw,
As if to prove saws knew what supper meant,
Leaped out at the boy's hand, or seemed to leap—
He must have given the hand. However it was,
Neither refused the meeting. (14–18)

The accident has happened before the reader has grasped what the narrator is saying. All this is quite deliberate on Frost's part for, after this first climax, the poem becomes increasingly terse as it builds to the second:

... They listened at his heart.
Little – less – nothing! – and that ended it.
No more to build on there. And they, since they
Were not the one dead, turned to their affairs. (31–34)

The ending of this poem has been read as a criticism of the callousness of those who can so easily turn their backs on tragedy and get on with their own lives; alternatively, it can be read as affirming the importance of life over death – an uncompromising realism rather than a sentimental grieving. The poem's ambiguities are strengthened by the use of an irregular half-rhyme all through the poem (*count/Vermont* [4, 6] *sister/ether* [26, 28] etc.). Significantly, the only end-word which is not given a half-rhyme is the word *hand* which rhymes with itself (18, 20); indeed, in the central four lines of the poem, the word *hand*, the focal image, appears four times.

It will be clear from this discussion that Frost's style is more carefully controlled than it may at first appear: far from just droning *on and on and on* (as Gibson recalled – see p.121), Frost as poet knows exactly how to pace and organize his narratives, and therefore how to modulate his 'voice'. What matters is that the reader is carried along by the rhythms of each sentence. These rhythms are pointed up by the repetitions, hesitations, qualifications, false-starts, and recapitulations that are some of the characteristic features of Frost's 'voice'. Again, '*Out, Out–*' (p.34) provides a good illustration:

> ... Then the boy saw all—
> Since he was old enough to know, big boy
> Doing a man's work, though a child at heart—
> He saw all spoiled. (22–25)

For Frost, the 'sound of sense' was the key to poetry. As early as 1912, when he arrived in England, his theory that the rhythms of a sentence were themselves an essential part of meaning dominated his thinking. It dominated his conversations with Edward Thomas, and found full expression in the poems of *North of Boston*. *I alone*, he boasted, *of English writers have consciously set out to make music out of what I may call the sound of sense.* (From a letter to John Bartlett, 4 July 1913.) Frost liked to draw attention to the sound of voices in the distance, or behind a closed door, when the intonation of the words could be heard, but not all the words themselves. Often it would be possible to infer the meaning, or at least the mood, of a sentence just from its sound. This is Frost's own formulation of his theory:

> The sound of sense ... is the abstract vitality of our speech. It is pure sound – pure form. One who concerns himself with it more than the sense is an artist. But remember we are still talking merely of the raw material of poetry. An ear and an appetite for these sounds of sense is the first qualification of a writer, be it of prose or verse. But if one is to be a poet he must learn to get cadences by skillfully breaking the sounds of sense with all their irregularity of accent across the regular beat of the metre. Verse in which there is nothing but the beat of the metre furnished by the accents of the polysyllabic words we call doggerel. Verse is not that. Neither is the sound of sense alone. It is a resultant from those two. (*Robert Frost: Poetry and Prose*, 1972, p.251)

Activity

Read *There Are Roughly Zones* (p.51). How clearly can Frost's 'voice', and the 'sound of sense', be heard in this poem?

Discussion

The poem is written in lines that are approximately equal: between ten and twelve syllables in mainly iambic metre. It has a rhyme scheme that begins conventionally (ABAB) but never regains any regularity until the end. The first line declares this to be another poem on the theme of inner and outer states of being: *We sit indoors and talk of the cold outside* (1). The speaker reports the thoughts of those inside the house and their conversation about the exotic peach tree threatened by the cold wind outside. The ebb and flow of their conversation is reflected in the parentheses that punctuate the otherwise smoothly-flowing lines – *we say* (5), *we admit* (6). The down-to-earth tenor of *There is nothing much we can do for the tree tonight* (14) echoes the familiar note of fatalism heard so often in Frost's poetry; on the other hand, when the speaker sees the planting of the tree in an alien climate as a symbol of man's desire to stretch himself to the limit, the tenor changes. The use of rhetorical question and the inversion of the word order achieves this effect:

> What comes over a man, is it soul or mind—
> That to no limits and bounds he can stay confined? (7–8)

In the final four lines of the poem, both tones are combined as Frost sums up the argument of the previous seventeen; he brings together the image of the tree and its metaphorical significance in a quatrain whose rhyme-scheme (ABBA) unites both elements:

> The tree has no leaves and may never have them again.
> We must wait till some months hence in the spring to know.
> But if it is destined never again to grow,
> It can blame this limitless trait in the hearts of men. (18–21)

Frost's Lyric Voice

By no means all Frost's writing is narrative, nor does he exclusively use the extended lines (ten syllables or longer) that some of his best known poems employ. On the other hand, he very rarely adopts free verse techniques, which he dismissed as *playing tennis without a net*. The discipline of form is important to him, and the variety of forms he uses reflects his versatility as a poet. *Mowing* (p.5) and *The Oven Bird* (p.29) are sonnets; *Acquainted with the Night* (p.46) is written in *terza rima* (three-line stanzas, where the middle line of one stanza provides the rhyming sound for the first and last lines of the next) while *To an Ancient* (p.58) is written in rhyming triplets. *An Unstamped Letter in Our Rural Letter Box* (p.56) and *For John F. Kennedy His Inauguration* (p.62) are written in rhyming couplets. Several poems (including *Stopping by Woods on a Snowy Evening*, p.41) are written in quatrains, while *The Road Not Taken* (p.28) has (uniquely) a five-line stanza rhyming ABAAB. In lyric poems such as these the voice of the speaker is often quieter and more introspective.

Nevertheless, such verse forms do not prevent Frost's 'voice' being recognized, though naturally a tighter scheme (particularly if it has much shorter lines) makes it unlikely that *the slow New England* fashion of speaking (see *The Golden Room*, p.121) will be so audible: in *The Sound of Trees* (p.35), for instance, Frost uses a seven-syllable line effectively interspersed with occasional lines of six syllables. The speaker thinks of the trees, with their constantly rustling branches, as being forever on the point of departure:

> They are that that talks of going
> But never gets away; (10–11)

Unlike the trees, however, the speaker himself is determined to go:

> I shall make the reckless choice
> Some day when they are in voice
> And tossing so as to scare
> The white clouds over them on.
> I shall have less to say,
> But I shall be gone. (20–25)

The number of syllables in each line is reduced as the poem ends, and in these six lines only three words have more than one syllable. The taut diction in a poem such as this creates an impressively emphatic note and reminds us how rarely in his poetry Frost uses long or unusual words at all.

When Frost uses long lines (typically decasyllabic or longer) he is generally in narrative mode. His shorter lines and lyric verse forms usually suggest that he is thinking aloud to himself rather than speaking to an audience. The 'voice' in such poems is often therefore muted, but the tone is no less audible.

Activity

Read *Neither Out Far nor In Deep* (p.50). How do you assess the tone of this poem?

Discussion

In this poem, statement and diction are pared to a minimum. Adjectives are rare (*one way* [2] *all day* [4]; *wetter ground* [7] *standing gull* [8]); there is only one simile (*like glass* [7]); the rhyme scheme and end-stopping give each line a strong, self-contained emphasis. The last stanza ends with a question to which there is no answer. Does the speaker have any sympathy with the people who still look outwards towards the horizon, even if there is not much to see? Does he wish he could join them or is he too cut off, too much an outsider, to be one of *the people* (12)? Alternatively, does he think their turning their backs on the land, the world they know, is a futile attempt to escape from the realities of life?

Ian Hamilton (*Robert Frost: Selected Poems*, p.21) describes the world evoked by this poem as:

> a friendless realm...It is difficult to imagine anything more terminally desolate than that. And yet the strain, the edginess, the intelligence of the poem derive really from a whole-hearted resistance to the terminal – a yearning for the conditions to be otherwise.

Another critic, Richard Poirier (*Robert Frost: The Work of Knowing*, p.269) argues that the landscape of the poem is:

> in every sense impoverished. It gives no sustenance to life; it promises little in the future, and none at all to the imagination ... the only hint of metaphoric activity in *Neither Out Far nor In Deep*, aside from the mockery in the title, is the observation that *The wetter ground like glass/Reflects a standing gull* (7–8). These lines, and others in the poem, emphasize the total *un*reflectiveness of *the people* (12) who merely sit all day and look at the sea. And what is further emphasized is the fact that no detail of the poem mirrors or reflects anything except inertia and conformity.

By contrast, an earlier critic of Frost, Randall Jarrell, argues that Frost sees the watchers in the poem as being foolish and yet heroic; both attitudes are implied, he suggests, because the last lines of the poem, in particular the concluding rhetorical question, contain a balance of tones which forbid outright condemnation or praise. Jarrell believes that the whole poem is *a recognition of the essential limitations of man, without denial, or protest or rhetoric or palliation.* ('To the Laodiceans', *Poetry and the Age*, p.39.)

Frost's most recent biographer, Jeffrey Meyers, reads the poem rather differently. He claims that:

> it mocks Frost's imperceptive critics, who turn their back on the reality of the land and look pointlessly at the sea all day. They cannot either look out far to see the whole design of his work, nor in deep to scrutinize the exact details ... The manifest limitations of his dull-witted critics, he says, never prevented them from searching for meanings in his verse, and their stupidity was never a bar to any watch they keep. This poem could be called, like the actual confection of ice cream encased in chocolate, 'Frost-Bite'. (*Robert Frost, A Biography*, p.215)

Conclusion

Such differences of opinion and interpretation as we have seen above should alert us to the danger of assuming that there is one 'right' way to read any poem, least of all one by Robert Frost. Frost's voice is distinctive, but it is not monotone; his view of the world is steady, but it is not blinkered. The landscapes that he inhabits and writes about are particular but his outlook is not insular. If, during his lifetime, some critics liked to attack him for not addressing the issues that seemed to them central to the age and its problems, Frost had a ready answer:

> Ages may vary a little. One may be a little worse than another. But it is not possible to get outside the age you are in to judge it exactly... Fortunately we don't need to know how bad the age is. There is something we can always be doing without reference to how good or bad the age is. There is at least so much good in the world that it admits of form and the making of form... When in doubt there is always form for us to be going on with. The artist, the poet, might be expected to be the most aware of such assurance. But it is really

everybody's sanity to feel it and live by it. (Quoted in *Robert Frost*, by Lawrance Thompson and R.H. Winnick, p.338)

Poetry, for Frost, is a way of giving form to life, of shaping experience and reflecting upon it. By inviting his readers to share in this enterprise (*You come too*) he still offers us the chance to hear one of the most uncompromising poetic voices of the Twentieth century:

Two roads diverged in a wood, and I—
I took the one less traveled by,
And that has made all the difference. (18–20)
(*The Road Not Taken*, p.28)

Chronology

1874 Robert Frost born in San Francisco
1885 Death of his father, William Frost; family moves to Massachusetts
1891 Meets his future wife, Elinor, at school
1892 Enrols at Dartmouth College, but leaves within a year
1894 First poem, *My Butterfly*, published in *The Independent*, New York
1895 Marries Elinor
1896 Birth of first son, Elliott
1897 Begins classes at Harvard University
1899 Leaves Harvard without completing his course; tries chicken farming
1900 Death of Elliott; death of his mother, Isabelle
1901 Frost's uncle helps him to purchase a farm at Derry, New Hampshire
1906 Begins teaching at Pinkerton Academy
1911 Sells the farm and begins teaching at Plymouth Normal School
1912 Sails to England with his family
1913 *A Boy's Will* published; meets Ezra Pound, W.B. Yeats, Georgian poets and Edward Thomas
1914 *North of Boston* published; moves to Dymock, Gloucestershire, with his family; outbreak of First World War
1915 Returns to USA; buys a farm in New Hampshire
1916 *Mountain Interval* published
1917 Begins teaching at Amherst College
1920 Leaves Amherst; moves to Vermont
1921 Takes up teaching post in Michigan
1923 Resigns teaching post in Michigan; returns to Amherst College; *Selected Poems* published; *New Hampshire* published
1924 Awarded Pulitzer Prize for *New Hampshire*
1925 Resigns from Amherst College; returns to Michigan
1926 Leaves Michigan and returns to Amherst (until 1938)
1928 Revisits Europe; renews acquaintance with the Dymock poets; *West-Running Brook* published

1930 *Collected Poems* published; wins second Pulitzer Prize
1934 Death of his youngest (and favourite) daughter, Marjorie, aged 29
1936 *A Further Range* published
1937 Awarded third Pulitzer Prize
1938 Death of his wife, Elinor
1939 Fellowship in Poetry at Harvard University (until 1941)
1940 Suicide of his son, Carol, aged 38
1942 *A Witness Tree* published; wins fourth Pulitzer Prize
1943 Fellowship at Dartmouth College (until 1949)
1947 *Steeple Bush* published
1949 Final return to Amherst College (until 1954)
1957 Visits England to receive honorary degrees at Durham, Oxford and Cambridge; lectures in London; final visit to Dymock
1961 Takes part in inauguration of President Kennedy
1962 *In the Clearing* published; visits Russia and meets Prime Minister Khrushchev as an American cultural ambassador just before the Cuban Missile Crisis
1963 Dies, 29 January

Further Reading

(Where a British and American edition exist, only the British has been listed.)

Editions

Edward Connery Lathem (ed.), *The Poetry of Robert Frost* (Jonathan Cape, 1971).

Edward Connery Lathem and Lawrance Thompson (eds.), *Robert Frost: Poetry and Prose* (Holt, Rinehart and Winston, Inc., New York, 1972).

Ian Hamilton (ed.), *Robert Frost: Selected Poems* (Penguin, 1973).

Richard Poirier and Mark Richardson (eds.), *Collected Poems, Plays and Prose by Robert Frost* (Library of America, 1995).

Lawrance Thompson (ed.), *Selected Letters of Robert Frost* (Holt, Rinehart and Winston, New York, 1964).

Biography

Eleanor Farjeon, *Edward Thomas: The Last Four Years* (Alan Sutton, revised ed., 1997).

Jeffrey Meyers, *Robert Frost, A Biography* (Constable, 1996).

William H. Pritchard, *Frost: A Literary Life Reconsidered* (University of Massachusetts Press, Amherst, 2nd. ed., 1993).

Lawrance Thompson and R.H. Winnick, *Robert Frost* (Holt, Rinehart and Winston, Inc., New York, 1981).

John Walsh, *Into My Own: The English Years of Robert Frost* (New York, 1988).

CD-Rom

Joe Matazzoni and David Sheehy (eds.), *Robert Frost: Poems, Life, Legacy* (Henry Holt, New York, 1997; 0-8050-5703-X). This interactive CD-Rom contains video and audio recordings of Frost, complete text of his poems, interviews, biographical and critical studies.

Criticism

Joseph Brodsky, Seamus Heaney, and Derek Walcott, *Homage to Robert Frost* (Faber and Faber, 1997).

Reginald Cook, *The Dimensions of Robert Frost* (Rinehart and Co., Inc., New York, 1958).

Randall Jarrell, 'To the Laodiceans' in *Poetry and the Age* (New York, 1955).

Judith Oster, *Toward Robert Frost: The Reader and The Poet* (University of Georgia Press, Georgia and London, 1991).

Richard Poirier, *Robert Frost: The Work of Knowing* (Stanford University Press, Stanford, California, revised ed. 1990).

Tasks

1 In what ways does *The Tuft of Flowers* (p.6) seem to you to be characteristic of Frost's themes and techniques?

2 Examine carefully the narrative technique adopted by Frost in *Mending Wall* (p.9). How clear is it to you what the final point of this poem is?

3 *After Apple-Picking* (p.25) is one of the most anthologized of all Frost's poems. Explore what seem to you to be the main reasons why it should be so popular. Is it, in your view, a good poem to represent Frost's particular themes and voice?

4 Read *The Wood-Pile* (p.26) carefully and choose two or three other poems from the selection with which it could be compared. Why do you think it has proved a less appealing poem than *After Apple-Picking*?

5 Examine carefully the diction (choice of words) employed by Frost in *The Cow in Apple Time* (p.32). Can you argue that the cow is for Frost a symbol of anything? Does the tone of the poem help to clarify Frost's attitude towards the cow?

6 Compare *Two Look at Two* (p.42) with *The Most of It* (p.52). What purpose does the intervention of the animals serve in each poem, and how effectively in your view does Frost present them in the two poems?

7 *In Hardwood Groves* (p.8), *Gathering Leaves* (p.44) and *A Leaf-Treader* (p.49) have significant similarities and differences. Explore these as a way of assessing the significance of leaves as images and metaphors in Frost's poetry.

8 *A Considerable Speck* (p.53) has been criticized for its apparent contempt for the modern world:

> I have none of the tenderer-than-thou
> Collectivistic regimenting love
> With which the modern world is being swept. (24–26)

Look carefully at the whole poem to decide what you think is Frost's main point, and whether you think the criticism is justified.

9 Compare the humour in *An Unstamped Letter in Our Rural Letter Box* (p.56) and *For John F. Kennedy His Inauguration* (p.62). Is the use of rhyming couplets equally effective in each poem?

10 In what ways could you call *The Middleness of the Road* (p.59) and *The Draft Horse* (p.66) typical Frost poems?

11 Look at the poems in which sleep and dreams play a part. How important is the world of the imagination and dreams for Frost compared with the world of experience?

12 All poetry benefits from being read aloud. In his public readings, Frost himself often had to read a poem more than once to get just the right intonation, stress or emphasis. Choose a range of four or five contrasting poems and, in a group, practise reading them aloud. In how many different ways can they be read? Does reading aloud help you to appreciate more clearly what Frost meant by 'the sound of sense'?

13 Look at the endings of Frost's poems. By choosing four or five (including perhaps *The Most of It*, p.52 and *The Oven Bird*, p.29) examine the way in which Frost changes or reinforces the apparent theme of a poem in the final lines.

14 Frost seems to me of vital interest and consequence because his ultimate subject is the interpretive process itself... His poetry is especially exciting when it makes of the 'obvious' something problematic, or when it lets us discover, by casual inflections or hesitations of movement, that the 'obvious' by nature *is* problematic. (Richard Poirier, *Robert Frost: The Work of Knowing*)

Use these comments as a starting point for an exploration of ways in which Frost presents the 'obvious' in his poetry.

15 Using the resources of libraries or multi-media, look up references to Frost in histories of modern American poetry and other reference works. By noting the dates of these references, can you detect any shift in the way critics have rated his importance during the past fifty years?

A publicity shot of Frost in 1916 taken for his collection *North of Boston*. Print from Amherst College Library

Facing page top
Portrait of Edward Thomas, Frost's close friend, taken in 1913

Facing page bottom
Frost with his wife Elinor at Plymouth in 1911

Frost at the inauguration of John F. Kennedy in 1961.
President Kennedy holds a book of new poems by Frost

Frost in old age with his dog

Appendix

Poems by and about Edward Thomas

The Sun Used to Shine

The sun used to shine while we two walked
Slowly together, paused and started
Again, and sometimes mused, sometimes talked
As either pleased, and cheerfully parted

Each night. We never disagreed
Which gate to rest on. The to be
And the late past we gave small heed.
We turned from men or poetry

To rumours of the war remote
10 Only till both stood disinclined
For aught but the yellow flavorous coat
Of an apple wasps had undermined;

Or a sentry of dark betonies,
The stateliest of small flowers on earth,
At the forest verge; or crocuses
Pale purple as if they had their birth

In sunless Hades fields. The war
Came back to mind with the moonrise
Which soldiers in the east afar
20 Beheld then. Nevertheless, our eyes

Could as well imagine the Crusades
Or Caesar's battles. Everything
To faintness like those rumours fades–
Like the brook's water glittering

Under the moonlight – like those walks
Now – like us two that took them, and
The fallen apples, all the talks
And silences – like memory's sand

When the tide covers it late or soon,
30 And other men through other flowers
In those fields under the same moon
Go talking and have easy hours.

Edward Thomas

As the Team's Head-Brass

As the team's head-brass flashed out on the turn
The lovers disappeared into the wood.
I sat among the boughs of the fallen elm
That strewed an angle of the fallow, and
Watched the plough narrowing a yellow square
Of charlock. Every time the horses turned
Instead of treading me down, the ploughman leaned
Upon the handles to say or ask a word,
About the weather, next about the war.
10 Scraping the share he faced towards the wood,
And screwed along the furrow till the brass flashed
Once more.
The blizzard felled the elm whose crest
I sat in, by a woodpecker's round hole,
The ploughman said. 'When will they take it away?'
'When the war's over.' So the talk began–
One minute and an interval of ten,
A minute more and the same interval.
'Have you been out?' 'No.' 'And don't want to, perhaps?'
'If I could only come back again, I should.
20 I could spare an arm. I shouldn't want to lose
A leg. If I should lose my head, why, so,
I should want nothing more . . . Have many gone
From here?' 'Yes.' 'Many lost?' 'Yes, a good few.
Only two teams work on the farm this year.
One of my mates is dead. The second day
In France they killed him. It was back in March,
The very night of the blizzard, too. Now if
He had stayed here we should have moved the tree.'
'And I should not have sat here. Everything
30 Would have been different. For it would have been
Another world.' 'Ay, and a better, though
If we could see all all might seem good.' Then

The lovers came out of the wood again:
The horses started and for the last time
I watched the clods crumble and topple over
After the ploughshare and the stumbling team.

Edward Thomas

Old Man

Old Man, or Lad's-love, – in the name there's nothing
To one that knows not Lad's-love, or Old Man,
The hoar-green feathery herb, almost a tree,
Growing with rosemary and lavender.
Even to one that knows it well, the names
Half decorate, half perplex, the thing it is:
At least, what that is clings not to the names
In spite of time. And yet I like the names.

The herb itself I like not, but for certain
10 I love it, as some day the child will love it
Who plucks a feather from the door-side bush
Whenever she goes in or out of the house.
Often she waits there, snipping the tips and shrivelling
The shreds at last on to the path, perhaps
Thinking, perhaps of nothing, till she sniffs
Her fingers and runs off. The bush is still
But half as tall as she, though it is as old;
So well she clips it. Not a word she says;
And I can only wonder how much hereafter
20 She will remember, with that bitter scent,
Of garden rows, and ancient damson trees
Topping a hedge, a bent path to a door,
A low thick bush beside the door, and me
Forbidding her to pick.
As for myself,
Where first I met the bitter scent is lost.
I, too, often shrivel the grey shreds,
Sniff them and think and sniff again and try
Once more to think what it is I am remembering,
Always in vain. I cannot like the scent,
30 Yet I would rather give up others more sweet,
With no meaning, than this bitter one.

I have mislaid the key. I sniff the spray
And think of nothing; I see and I hear nothing;
Yet seem, too, to be listening, lying in wait
For what I should, yet never can, remember:
No garden appears, no path, no hoar-green bush
Of Lad's-love, or Old Man, no child beside,
Neither father nor mother, nor any playmate;
Only an avenue, dark, nameless, without end.

Edward Thomas

Wind and Mist

They met inside the gateway that gives the view,
A hollow land as vast as heaven. 'It is
A pleasant day, sir.' 'A very pleasant day.'
'And what a view here! If you like angled fields
Of grass and grain bounded by oak and thorn,
Here is a league. Had we with Germany
To play upon this board it could not be
More dear than April has made it with a smile.
The fields beyond that league close in together
10 And merge, even as our days into the past,
Into one wood that has a shining pane
Of water. Then the hills of the horizon–
That is how I should make hills had I to show
One who would never see them what hills were like.'
'Yes. Sixty miles of South Downs at one glance.
Sometimes a man feels proud of them, as if
He had just created them with one mighty thought.'
'That house, though modern, could not be better planned
For its position. I never liked a new
20 House better. Could you tell me who lives in it?'
'No one.' 'Ah – and I was peopling all
Those windows on the south with happy eyes,
The terrace under them with happy feet;
Girls – ' 'Sir, I know. I know. I have seen that house
Through mist look lovely as a castle in Spain,
And airier. I have thought: ''Twere happy there
To live.' And I have laughed at that
Because I lived there then.' 'Extraordinary.'
'Yes, with my furniture and family
30 Still in it, I, knowing every nook of it
And loving none, and in fact hating it.'
'Dear me! How could that be? But pardon me.'
'No offence. Doubtless the house was not to blame,

But the eye watching from those windows saw,
Many a day, day after day, mist–mist
Like chaos surging back – and felt itself
Alone in all the world, marooned alone.
We lived in clouds, on a cliff's edge almost
(You see), and if clouds went, the visible earth
40 Lay too far off beneath and like a cloud.
I did not know it was the earth I loved
Until I tried to live there in the clouds
And the earth turned to cloud.' 'You had a garden
Of flint and clay, too.' 'True; that was real enough.
The flint was the one crop that never failed.
The clay first broke my heart, and then my back;
And the back heals not. There were other things
Real, too. In that room at the gable a child
Was born while the wind chilled a summer dawn:
50 Never looked grey mind on a greyer one
Than when the child's cry broke above the groans.'
'I hope they were both spared.' 'They were. Oh yes!
But flint and clay and childbirth were too real
For this cloud-castle. I had forgot the wind.
Pray do not let me get on to the wind.
You would not understand about the wind.
It is my subject, and compared with me
Those who have always lived on the firm ground
Are quite unreal in this matter of the wind.
60 There were whole days and nights when the wind and I
Between us shared the world, and the wind ruled
And I obeyed it and forgot the mist.
My past and the past of the world were in the wind.
Now you may say that though you understand
And feel for me, and so on, you yourself
Would find it different. You are all like that
If once you stand here free from wind and mist:
I might as well be talking to wind and mist.

You would believe the house-agent's young man
70 Who gives no heed to anything I say.
Good morning. But one word. I want to admit
That I would try the house once more, if I could;
As I should like to try being young again.'

Edward Thomas

Easter Monday

(In Memoriam E.T.)

In the last letter that I had from France
You thanked me for the silver Easter egg
Which I had hidden in the box of apples
You liked to munch beyond all other fruit.
You found the egg the Monday before Easter,
And said, 'I will praise Easter Monday now–
It was such a lovely morning.' Then you spoke
Of the coming battle and said, 'This is the eve.
Good-bye. And may I have a letter soon.'

10 That Easter Monday was a day for praise,
It was such a lovely morning. In our garden
We sowed our earliest seeds, and in the orchard
The apple-bud was ripe. It was the eve.
There are three letters that you will not get.

Eleanor Farjeon

Index of Titles and First Lines

Cork City Library
WITHDRAWN
FROM STOCK